Environmental Impact of Power Generation

ISSUES IN ENVIRONMENTAL SCIENCE AND TECHNOLOGY

TITLES IN THE SERIES:

1 Mining and its Environmental Impact
2 Waste Incineration and the Environment
3 Waste Treatment and Disposal
4 Volatile Organic Compounds in the Atmosphere
5 Agricultural Chemicals and the Environment
6 Chlorinated Organic Micropollutants
7 Contaminated Land and its Reclamation
8 Air Quality Management
9 Risk Assessment and Risk Management
10 Air Pollution and Health
11 Environmental Impact of Power Generation

FORTHCOMING:

12 Endocrine Disrupting Chemicals

How to obtain future titles on publication

A subscription is available for this series. This will bring delivery of each new volume immediately upon publication. For further information, please write to:

The Royal Society of Chemistry
Turpin Distribution Services Limited
Blackhorse Road
Letchworth
Herts SG6 1HN, UK

Telephone: +44 (0) 1462 672555   Fax: +44 (0) 1462 480947

ISSUES IN ENVIRONMENTAL SCIENCE
AND TECHNOLOGY

EDITORS: R. E. HESTER AND R. M. HARRISON

11

Environmental Impact of Power Generation

ROYAL SOCIETY OF CHEMISTRY

ISBN 0-85404-250-4
ISSN 1350-7583

A catalogue record for this book is available from the British Library

Published by The Royal Society of Chemistry, Thomas Graham House,
Science Park, Milton Road, Cambridge CB4 0WF, UK
For further information see our web site at www.rsc.org

Typeset in Great Britain by Vision Typesetting, Manchester
Printed and bound by Redwood Books Ltd., Trowbridge, Wiltshire

Preface

A popular image of the power generation industry is one of chimney stacks belching out fumes from the combustion of fossil fuels, causing pollution of the atmosphere and contamination of land and water. To what extent has this poor image been justified in the past, and what is happening today to improve matters? What performance standards does the industry achieve? And what about effects such as those associated with high voltage overhead power lines, and worries about nuclear power stations? Answers to these and other such questions are provided by this book.

In the first article, Gordon MacKerron, Head of the Energy Programme at SPRU in the University of Sussex, provides an historical context and overview of the electricity supply industry. This suggests that as environmental regulation has become more stringent and more integrated over time, so in the areas of acid rain and climate change the electricity generating sector has been increasingly required to play a role disproportionate to its damage contribution in the solution of problems. The second article, by Bernard Fisher, Professor of Environmental Modelling in the School of Earth and Environmental Sciences at the University of Greenwich, addresses issues related to the airborne emissions from power stations over the past 40 years. This includes the road transport sector which has come under greater scrutiny in recent years. It concludes that recently adopted air quality management approaches are to be recommended because they permit all categories of sources to be judged on a common basis.

A view from the UK power industry is provided next in an article by Stephen Adrain and Ian Housley, both of National Power plc. They discuss the evolution of the industry towards sustainability through reductions in emissions, environmental impacts, and use of resources within the context of a liberalised market. They point to the fact that the industry has adopted far reaching policies and implementation strategies which are resulting in major environmental benefits. Complementing this article, the following one by Colin Powlesland of the UK government Environment Agency, discusses the application of the Best Practicable Environmental Option (BPEO) approach to design and siting of power stations. Key stages in this approach are the definition of study objectives, data collection, selection of options for assessment, environmental and economic assessment, selection of the preferred option and its presentation. Running through this methodology is the need to maintain an audit trail.

In the fifth article of the book the environmental impact of the nuclear fuel cycle, from ore to power generation, is explored by Simon Port of British Nuclear Fuels Ltd. (BNFL) and the University of Central Lancashire together with Malcolm Joyce of the University of Lancaster. They describe and comment on both radiological and non-radiological factors that influence public awareness, economics, and safety in this sector of the power industry. The complete nuclear fuel cycle is described and placed in context with the use of alternative energy sources along with the risks associated with their operation. The radiological impact is compared directly with the background levels to enable the reader to assess its significance.

Electromagnetic fields and possible ecological effects of overhead transmission lines are the subject of the penultimate article, written by David Jeffers who is an EMF Consultant to the National Grid Company plc. The impact of the electrical activity on the conductor surface on the generation of ozone, oxides of nitrogen, and ions is discussed and the results of long-term monitoring studies are described. Electrical fields have been found to have adverse effects on bees and tree growth, but studies of their impact on farm animals and crops have shown no such effects. Guidelines on the limitation of exposure to electric and magnetic fields are summarised, together with conclusions from recently published reviews of epidemiological data. The final article is by Andrew Warren, Director of the Association for the Conservation of Energy. His theme is energy efficiency and conservation and includes an introduction to the concept of 'least cost planning' and the related activities of 'integrated resource planning' and 'rational planning' as used by energy utilities in the USA and Europe. Issues in demand-side management are considered together with the costs and benefits of energy efficiency programmes and the valuation of environmental damage.

We believe this wide-ranging treatment of the many issues associated with power generation and its environmental impact will be found useful both within and outside the industry. It will make a valuable contribution to the public understanding of science in this important area and should be essential reading for students in many engineering and environmental science courses.

Ronald E. Hester
Roy M. Harrison

Contents

Issues in Environmental Science and Technology, No. 11
Environmental Impact of Power Generation
© The Royal Society of Chemistry, 1999

Contents

Editors

Ronald E. Hester, BSc, DSc(London), PhD(Cornell), FRSC, CChem

Ronald E. Hester is Professor of Chemistry in the University of York. He was for short periods a research fellow in Cambridge and an assistant professor at Cornell before being appointed to a lectureship in chemistry in York in 1965. He has been a full professor in York since 1983. His more than 300 publications are mainly in the area of vibrational spectroscopy, latterly focusing on time-resolved studies of photoreaction intermediates and on biomolecular systems in solution. He is active in environmental chemistry and is a founder member and former chairman of the Environment Group of The Royal Society of Chemistry and editor of 'Industry and the Environment in Perspective' (RSC, 1983) and 'Understanding Our Environment' (RSC, 1986). As a member of the Council of the UK Science and Engineering Research Council and several of its sub-committees, panels and boards, he has been heavily involved in national science policy and administration. He was, from 1991–93, a member of the UK Department of the Environment Advisory Committee on Hazardous Substances and is currently a member of the Publications and Information Board of The Royal Society of Chemistry.

Roy M. Harrison, BSc, PhD, DSc (Birmingham), FRSC, CChem, FRMetS, FRSH

Roy M. Harrison is Queen Elizabeth II Birmingham Centenary Professor of Environmental Health in the University of Birmingham. He was previously Lecturer in Environmental Sciences at the University of Lancaster and Reader and Director of the Institute of Aerosol Science at the University of Essex. His more than 250 publications are mainly in the field of environmental chemistry, although his current work includes studies of human health impacts of atmospheric pollutants as well as research into the chemistry of pollution phenomena. He is a past Chairman of the Environment Group of The Royal Society of Chemistry for whom he has edited 'Pollution: Causes, Effects and Control' (RSC, 1983; Third Edition, 1996) and 'Understanding our Environment: An Introduction to Environmental Chemistry and Pollution' (RSC, Second Edition, 1992). He has a close interest in scientific and policy aspects of air pollution, having been Chairman of the Department of Environment Quality of Urban Air Review Group as well as currently being a member of the DETR Expert Panel on Air Quality Standards and Photochemical Oxidants Review Group, the Department of Health Committee on the Medical Effects of Air Pollutants and Chair of the DETR Atmospheric Particles Expert Group.

Contributors

R.S. Adrain, *National Power plc, Windmill Hill Business Park, Whitehill Way, Swindon SN5 6PB, UK*

B.E.A. Fisher, *School of Earth and Environmental Sciences, University of Greenwich, Chatham ME4 4TB, UK*

I. Housley, *National Power plc, Windmill Hill Business Park, Whitehill Way, Swindon SN5 6PB, UK*

D. Jeffers, *National Grid Company, Brookmead, Guildford Business Park, Middleton Road, Guildford GU2 5XQ, UK*

M.J. Joyce, *Department of Engineering, University of Lancaster, Bailrigg, Lancaster LA1 4YR, UK*

G. MacKerron, *SPRU, University of Sussex, Mantell Building, Falmer, Brighton BN1 9RF, UK*

S.N. Port, *Centre for Materials Research, University of Central Lancashire, Preston PR1 2HE, UK*

C. Powlesland, *National Centre for Risk Analysis and Options Appraisal, Environment Agency, Steel House, 11 Tothill Street, London SW1H 9NF, UK*

A. Warren, *Association for the Conservation of Energy, Westgate House, Prebend Street, London N1 8PT, UK*

Historical Overview

GORDON MacKERRON

1 Introduction

Around the turn of the last century, a great wave of technological innovations transformed industrial and domestic life in the industrialized countries. These innovations included the motor car and the modern chemical industry, but an essential ingredient in the transformation was the provision of electricity supply, with the steam turbine playing a vital role. The advantages of electricity at the point of use have always been that it is clean, precise, and efficient. However, as the 20th century progressed, it became clear that to *generate* electricity cheaply it was necessary to move to larger and larger scales, and often to sites that were remote from the main centres of electricity demand. This, in turn, brought a need for long-distance transmission links which changed the physical appearance of parts of the countryside. As concern with various aspects of air quality grew strongly in the second half of the century, so power generation became increasingly implicated in the major issues: particulates, sulfur, nitrogen, and most recently carbon, besides the emotive issues surrounding nuclear power.

By the late 20th century, the electricity industry had thereby become deeply enmeshed in most of the leading environmental problems of concern to both Governments and citizens. Almost all major forms of electricity generation—fossil fuel-based, nuclear, large hydro, newer renewables, as well as transmission—have raised serious environmental concerns. This chapter, in keeping with the rest of the book, concentrates on the issues that are specific to the UK, and therefore gives little consideration to environmental concerns surrounding large hydro schemes. However, for virtually all countries, the impacts of electricity generation are high on the list of active environmental issues, and many of these issues are now subject to international and even global negotiation and control.

2 History

Electricity supply in the form of public lighting stretches back to the early 1880s, with Godalming in Surrey and Brighton having the earliest public supply

Issues in Environmental Science and Technology, No. 11
Environmental Impact of Power Generation

systems. However, gas remained a powerful lighting competitor for many decades and the electric lighting schemes remained mostly very small. The development of trams and railway electrification represented a major growth in the use of electricity, and by the first decade of the 20th century, electricity had begun to be an important source of motive power for industry. By the time of the First World War, factory power had overtaken traction and lighting in terms of kilowatt hours used.[1] Most of the 19th century uses of electricity depended on small, on-site forms of generation, often using reciprocating engines. As the steam turbine—with its potential for efficiency on a larger scale—became more widely used, so 'central' power stations with local distribution networks became more common. Newcastle was a leader in this, and the Newcastle Electricity Supply Company operated the largest integrated power system in Europe before 1914. However, the industry remained small and localized before 1914, and its environmental impact was small.

The First World War accelerated the development of interconnected and larger systems, and the potential for household electricity use began to be exploited in the post-war period. The potential benefits of large scale interconnection were beginning to be clear, but a major problem was the huge variety of systems, both municipal and private, that were developed on a local scale. Standardization and rationalization were both necessary and difficult, but the watershed was the setting up of a Central Electricity Board under the Electricity Supply Act of 1926. Its main task was the establishment of a national grid system. This was the first time that electricity impinged on rural environments, and great dispute surrounded the intrusion of overhead pylons and wires into areas of natural beauty. The first 'amenity' based pressure groups began to be set up in the 1920s—for example, the still-active Council for the Protection of Rural England—and electricity transmission was one of the issues that engaged this amenity movement. By the mid-1930s, the grid was virtually complete and the efficiency and cost-reducing benefits were large, not least because the need to keep large reserve margins of generating capacity fell sharply.

In the early days of grid operation, the main purpose was to connect the main industrial areas and provide back-up, and siting of power stations remained essentially urban and close to main load centres. This meant that, in emission terms, it was urban areas which suffered most, and the principal problem was smoke or particulates emerging from the chimneys of the almost exclusively coal-fired urban stations. However, there were so many other sources of local air pollution—urban factories and, in the winter, private homes—that the contribution of power generation was not especially large or noticeable in the period up to the Second World War.

It was not until the Second World War and afterwards that the grid began to operate as a truly national, rather than a regionally based, system. In 1947, the UK industry was nationalized, and much rationalization remained to be done. Nationalization brought together 200 companies, 369 local authority undertakings, nearly 300 power stations, and the Central Electricity Board under the new British Electricity Authority (BEA).[2] Within this new structure, 14 Area Boards

---

[1] I. C. R. Byatt *The British Electrical Industry 1875–1914*, Clarendon Press, Oxford, 1979.
[2] L. Hannah, *Engineers, Managers and Politicians*, Macmillan Press, London, 1982.

2

were to be responsible for regional distribution, and a new Central Authority (later to split into the CEGB for England and Wales, and two integrated Boards for Scotland) took over power stations and high voltage transmission.

Under the new order, standardization was completed, and from the 1950s onwards the size of generating sets increased rapidly from 30 and 60 megawatts (MW) to 660 MW in the 1970s, by which time 2000 MW stations were normal. This vast increase in scale was accompanied by a radical shift in siting policy. Urban sites, except for a few gas turbines in the 1960s and 1970s, were no longer used for new investment. Coal was the dominant fuel, especially in the 1950s, and here the practice was to locate power stations on the coal-fields. As oil became important in the 1960s, oil-fired stations were built at coastal sites, usually near oil refineries, and at the same time nuclear stations began to be sited in remote areas, nearly always at a coastal location. This all involved larger flows of power over longer distances, but the development of the super-grid at 275 kilovolts (kV) and later 400 kV meant that transport costs fell sharply. Because of the accidents of coal-field locations, the dominant direction of power flow in England was from the north and north-west to the south and south-west.

Electricity demand grew rapidly in the 1950s and 1960s, commonly at around 7% annually. Despite increases in thermal efficiency, this inevitably meant large increases in power station emissions, though the policy of tall stacks and the fitting of electrostatic precipitators meant that the power sector was not heavily involved in the growing air pollution issues of this period, notably the very poor urban air quality that led to the passing of the 1956 Clean Air Act.[3] Ironically, just as electricity demand growth faltered in the wake of the oil crises and economic recessions of the 1970s, so environmental pressures on the industry began to build up. These pressures—which are the main subject of the rest of the chapter—were mainly to do with different forms of emissions and air pollution, as well as the particular problems of nuclear power.

3 The Nature and Importance of Environmental Impacts

Environmental impacts are wide-ranging and various in many dimensions. Physically, both the medium (air, land, water) and extent of geographical area vary widely. Politically, environmental issues excite controversy to variable and changing extents, and are dealt with at all political levels, from parish council to global inter-Governmental negotiation.

Because of the diversity of issues, it is impossible to find any satisfactory way of dealing with all environmental issues using a common measuring device, despite major attempts by economists to use money in this role in recent years. There are two reasons for the difficulty of expressing environmental harm along one dimension:

- First, there is the physical science problem: our knowledge of the extent of physical damage created by different forms of environmental impact is

[3] L. E. J. Roberts, P. S. Liss and P. A. H. Saunders, *Power Generation and the Environment*, Oxford University Press, Oxford, 1990.

incomplete and uncertain (especially as we move to the larger and global questions like climate change)

● Second, there is no obvious way of directly comparing the damage caused by, say, visual intrusion, sulfur dioxide emissions, and radioactive waste. This is partly because the kinds of harm created are so different in these cases, and more importantly because the attempt by economists to provide monetary values for many kinds of damage have been controversial and unsatisfactory.[4] The underlying problem is that there is nothing remotely resembling a market in which many forms of environmental damage could be valued. The attempts by economists to infer market values from other markets (hedonic valuation) and by asking people directly what would be their willingness to pay to avoid damage (contingent valuation) produce results that are often inconsistent and somewhat forced.[5] Despite very large and sophisticated projects which have attempted to find consistent and useful valuations of environmental damage,[6] these techniques do not provide satisfactory answers. There is, therefore, no escape from the need for Governments and citizens to make their own judgements about the relative importance of different environmental issues in framing policy responses

Despite these difficulties in comparing different forms of environmental impact, it is useful, in looking at the impacts made by the electricity supply industry (and more widely), to think in terms of a broad three-fold classification. This has geographical, historical, physical, and scientific dimensions:

● The earliest form of modern environmental concern, dating from the 1920s, concentrated on local and highly tangible or visible impacts. The classic inter-war issues were about siting of industrial facilities in rural environments and, as mentioned earlier, the major issue for electricity was the overhead transmission lines that so rapidly spread across the country in the decade after 1925. This was a concern about aesthetics and amenity. The impact itself was visible and generally did not need the mediation of science to be apparent, but the geographical scope of impact was limited to a few miles of the source of damage. Local concerns of this sort remain prominent and the 'pylons' disputes of 70 years ago have recent echoes in the disputes about the siting of wind turbines

● In the period between the 1950s and the early 1980s there emerged a set of concerns about environmental damage with increasingly wider geographical impacts and needing more sophisticated scientific mediation. For electricity these issues were almost entirely about 'air pollution' of various sorts. In the 1950s this was mainly a matter of local air quality (smoke and smogs) in which electricity played a relatively small part, but the scope of perceived problems grew substantially in the 1960s and 1970s, and the classic issue became acidification—long-distance transport of sulfur and nitrogen and

[4] M. Jacobs, *The Green Economy*, Pluto Press, London, 1991.

[5] D. Pearce, *Blueprint for a Green Economy*, Earthscan, London, 1989.

[6] European Commission, *Externe: Externalities of Energy*, EUR 16520-16525 EN, Brussels, 1995, vols. 1–6.

their impact on lakes, forests and land. Here, scientific work became crucial to the definition of causes and, while the impacts were still tangible, the connections between cause and effect were becoming less clear-cut. The geographical scope became national and even international. In the UK, sulfur became a serious issue only in the 1980s, but in the USA and Japan it had emerged as a serious problem by the 1970s

- From the 1970s to the present, the new issues have become even wider. The classic issues are now potentially continental and global in scope. For electricity, the two major issues in this category are the impact of radioactive releases from nuclear power stations or waste facilities, and the issue of climate change. In the nuclear case the potential was clear from the 1950s, but it was not until the Chernobyl accident of 1986 that the geographical extent of the resulting problems became fully apparent. For these widest of all issues, the potential impacts are very large indeed, but are entirely obscure to public perception without the mediation of sophisticated science. The role of science in the definition of damage to human health from radiation, and in predicting the extent and nature of climate change, are critical to the debates. A substance like carbon dioxide is only definable as a 'pollutant' under particular scientific assumptions about the effects of concentrations of gases in the atmosphere. Solutions to this kind of environmental issue depend on very complex inter-Governmental negotiation.

In the second half of the 20th century, environmental issues have become continuously more important at a political level and this is reflected in ever-increasing stringency in environmental regulation. There seem to be three types of explanation for this well-recognized phenomenon of growing stringency:

- As industrial activity has expanded further, there have been large increases in the production of well-known pollutants like sulfur dioxide, and correspondingly greater physical damage
- New work in the environmentally related sciences has often led to the discovery of new forms of damage not previously understood (for example, damage to human health from exposure to lead in petrol)
- Changes in culture and society seem to lead to more concern about avoiding or mitigating *given* levels of damage. In more economic language, increased wealth in the industrialized countries has been associated with a sharply increasing demand for better environmental quality, expressed both politically and, to some extent, in the market place. There are probably other elements involved here. Some newer kinds of potential environmental harm—classically, ionizing radiation and, increasingly, genetic modification—involve new and powerful interventions by science into the natural world, with less immediately tangible but potentially apparently catastrophic impacts. It seems likely that these new technology developments have added an extra twist to this demand for better environmental quality, as well as providing a fresh dimension of political controversy.[7]

[7] T. O'Riordan, *Environmentalism*, Pion, London, 1981.

The simultaneous effect of these three connected but in some ways distinct phenomena has proved a very powerful force in raising environmental issues high up the political agenda of the industrialized world. The apparently inexorable increases in industrial output and consequent environmental damage, added to scientific confirmation that given levels of emission or pollution are more harmful than previously thought, combined with new and potentially devastating risks like releases of radioactivity, seem to have brought a powerful change in personal and political attitudes to environmental harm.

4   Environmental Regulation: the Framework

Early History

Environmental regulation in the UK stretches further back than the origins of the electricity supply industry. In the air pollution domain, the Alkali Inspectorate was set up in 1863 to deal with large-scale emissions, largely of hydrochloric acid from the chemicals industry, that were ruining large areas of agricultural land. The Inspectorate was intended to be separate from local authorities, on which factory owners were often influential, but from the earliest days the preferred method of working of the Inspectorate was co-operative rather than confrontational, a tradition that has persisted to the present day. The first Alkali Inspector, Robert Smith, coined the term 'acid rain'.

Emission limits were prescribed for the first time in 1874, and the 1906 Alkali &c Works Regulation Act set the pattern of environmental regulation for much of the 20th century. The Act specified lists of 'noxious and offensive gases' and a similar list of industrial processes. It also introduced the notion of 'best practicable means' (BPM) as a guiding principle for regulating emissions where there were no statutory limits. At a different and more local environmental level, the Electricity Act of 1909 began the process of controlling the siting of power stations as a land use issue.

The next major developments in environmental control followed the great smogs of the early 1950s in London with the passage of the Clean Air Act of 1956. This, as mentioned earlier, had little direct impact on the electricity supply industry. While electricity generation was still principally an urban activity, policies to encourage 'tall stacks' and fitting of electrostatic precipitators to remove particulates meant that the contribution of power stations to urban smogs was relatively small. The Clean Air Act, in addition, did not require the fitting of FGD or other sulfur-removal systems to existing or new power stations. However, the tall stacks policy, while reducing local emission problems, also did much to help deposit emissions much further afield, including internationally.

In the following year, 1957, the Electricity Act bound generators to a wide range of environmental responsibilities at the site level and its immediate surroundings. Provisions in the Act are in the broad area of aesthetics and amenity and require all parties to protect the beauty, flora, fauna, buildings, and other objects on which power stations might have an impact. Following this Act, the newly established CEGB set up a Station Environment Group to help integrate environmental issues into station planning.

Nuclear power began to be commercialized in the 1960s, and various arrangements were made to reflect the particular problems of controlling radiation. There had always been a longer tradition of international debate and policy making in nuclear power, stretching back to 1928 when the body later named the International Commission on Radiological Protection (ICRP) was formed. A Nuclear Installations Inspectorate (NII) was created in the UK to issue site licences and exercise control over design and safety practice at nuclear plants, though regulation of substantial parts of the nuclear power industry, outside the Generating Boards, was exercised internally by the UK Atomic Energy Authority (UKAEA). The Department of the Environment and the Ministry of Agriculture, Fisheries and Food are also responsible for granting discharge authorizations for specified amounts of routine radioactive releases. A complex framework of other bodies contributed to nuclear regulation and advice, including the National Radiological Protection Board (NRPB), which advised Government on questions of radiological protection, and later the Radioactive Waste Management Committee (RWMAC), which advises more widely on policy questions. The guiding principle in nuclear regulation was that doses of radiation should be made As Low as Reasonably Achievable (ALARA), which was broadly equivalent in the nuclear field to the BPM idea in other areas of pollution control.

From the 1970s to the Present

Until the 1970s the tradition in UK environmental regulation involved a piecemeal approach, with different inspectorates, largely independent of each other, regulating the different media of land, air, water, and nuclear power. 'Environmentalism' had not yet developed to any great extent, and little need had been seen to have a general environmental approach to issues, let alone a strategy. However, there were criticisms of the closed nature of the decision making by the pollution inspectorates: the close working relationships between polluters and inspectors encouraged the idea of regulatory capture, and regulation was site-specific and largely implemented on a voluntary basis.[8] Large changes began in the 1970s and two in particular may be mentioned:

- Attempts began to be made to integrate the different forms of environmental regulation under a common body with a more consistent approach
- From the early 1980s, environmental policy stopped being a purely domestic affair, and would henceforth need to take account of an important European (and later global) dimension

Under the 'integration' heading, the first major move came in 1974. Under the Health and Safety at Work Act, the first attempt at integration involved trying to look across a wide range of risk issues (worker health and safety as well as emissions) under a single body, the Health and Safety Commission (HSC). The Alkali Inspectorate (renamed the Industrial Air Pollution Inspectorate, or IAPA,

---

[8] J. Skea and A. Smith, in *Britain in Europe: National Environmental Policy in Transition*, ed. P. Lowe, Routledge, London, 1996.

in 1982) together with the NII came under the overall control of the HSC. This attempt to merge worker safety and environmental issues did not work well, and in practice the HSC provided only a very loose federal layer over a number of bodies. The environmental inspectorates still operated in an independent way and did not merge with the Factories Inspectorate.

The 1974 changes were administrative rather than substantive—there was no real attempt at integration across environmental policy and regulation. In 1976 the Royal Commission on Environmental Pollution (RCEP), responding to concerns about the closed and piecemeal approach to environmental regulation, recommended that there should be a unified pollution inspectorate, with publicly available and legally binding emission limits.[9] It also believed that environmental regulation should be under the control of the Department of the Environment (rather than the Department of Employment, to which the HSC was ultimately responsible). The 1976 report of the Royal Commission was a landmark in introducing notions of integrated pollution control and in recommending more open and accountable procedures.

However, the response of the Government to the RCEP was slow, and it took a further 10 years (1986) before a major review of air pollution was undertaken. In 1987, the Department of the Environment, which had spent much energy in the 1980s raising the profile of environmental issues to a public increasingly receptive to environmentalism, succeeded in gaining control over environmental regulation. A new body, Her Majesty's Inspectorate of Pollution (HMIP), was created under the Department. The IAPI was brought together with other regulatory offices, including the Radiochemical Inspectorate and a new water inspectorate.

The creation of HMIP also foreshadowed the introduction of Integrated Pollution Control (IPC), and in 1990 an Environmental Protection Act gave the necessary legislative backing to the implementation of the idea of IPC. The Act was comprehensive and a landmark in environmental regulation. Besides introducing the legislative backing for IPC, it introduced a new principle to replace BPM. This was Best Available Technique Not Entailing Excessive Cost (BATNEEC), which is recognizably related to BPM, but which gives more explicit attention to the economic dimension of the decision process. In addition, the principle of the Best Practicable Environmental Option (BPEO) had to be considered for processes where cross-media environmental impacts were likely. This was a practical implementation of the idea of integrating pollution control across media. In addition, provisions were made, for the first time, for public access to regulatory information and emission monitoring data.

The attempt at integration across national environmental policy in the 1990 Environment Protection Act was continued in the early 1990s. In 1995, there was a new Environment Act, through which the functions of HMIP, waste regulation authorities, and the National Rivers Authority were combined in a single Environment Agency, which began operation in 1996. For the first time in the UK, virtually all environmental regulation had been brought together under a single agency.

---

[9] Royal Commission on Environmental Pollution, *Air Pollution: an Integrated Approach*, 5th Report, HMSO, London, 1976.

The second major change after the 1970s was that environmental regulation stopped being a purely national affair. There were essentially two causes of the growing 1980s trend towards Europe-wide (and later global) environmental regulation. The first of these was the obvious fact that many environmental issues were of a transboundary character, and could only be tackled effectively through international co-operation. The second was a political factor separate from the environment: in the push towards closer integration in the European Community, it became increasingly important to establish a 'level playing field' so that freer trade between Member States could be seen to be fair. Widely differing environmental standards were clearly a breach of this idea, and Germany (with relatively stringent domestic environmental regulation) pushed particularly hard to implement the level environmental playing field.[10] Early results were the European Directives on Air Pollution of 1984, and the much-debated Large Combustion Plant Directive (LCPD) of 1988. One of the many purposes of the 1990 UK Act was to provide a formal basis for national implementation of such Directives

From the 1970s onwards, the transboundary character of many environmental issues meant that international negotiations and agreements were increasingly important in setting frameworks of policy and national emission targets. Together with the Economic Commission for Europe (ECE), the European Community was the main forum for this increasing trans-nationalization of environmental policy and Member States were, in the 1980s, increasingly required to find ways of translating Europe-wide Directives into national practice. Of particular relevance to the ESI were the 1984 European Directive on Air Pollution from Industrial Plants and the 1988 Large Combustion Plant Directive, which set national limits on future sulfur dioxide emissions. In 1995 the European Union agreed an Integrated Pollution Prevention and Control Directive which aimed at IPC in a similar fashion to that already being implemented in the UK, though its scope was somewhat wider.

Meanwhile, an even wider international stage was becoming important in relation to climate change issues. The Rio 'Earth Summit' of 1992 was the first major step in the attempt to bring some international agreement over the greenhouse gases (GHGs) and this was followed by the Kyoto meeting of December 1997, in which the agreements reached will, for the first time in the GHG sphere, be legally binding.

Two further developments in recent years are worth remarking on. The first is that the UK Government has become interested in the attempt to introduce market-based forms of environmental control to supplement the traditional 'command-and-control' physically based standards. These market-based instruments could be either in the form of 'green taxes' or (increasingly favoured in recent years) tradable emission permits. Practical progress with such 'economic' instruments has been limited—only a small landfill tax has yet been implemented—but the tradable permit issue seems likely to become a reality as methods are worked out to implement the Kyoto agreement on greenhouse gases.

The other recent development is the gradual broadening of the overall

---

[10] S. Boehmer-Christiansen and J. Skea, *Acid Politics*, Belhaven, London, 1991.

philosophy of environmental control. From the 1970s the UK, along with the rest of the OECD, subscribed to the idea of the 'polluter pays', an idea designed to ensure that the often external costs of damage were borne by those who created the damage. In the 1980s the idea of the precautionary principle became more important: this was the notion that where scientific uncertainty exists about damage, precaution demands that policy is framed on the basis that such potential damage be treated as having a high probability. By 1991, the UK Government officially embraced the even wider idea of 'sustainability' as the overarching principle of environmental policy.[11] Polluters paying and the precautionary principle can be seen as subsidiary objectives within the sustainability idea.

The notion of sustainability is imprecise, but is summarized in the idea that the present generation, in meeting its needs, should not jeopardize the ability of future generations to meet their needs.[12] The implications of the application of this principle to any particular area of environmental policy or regulation need to be established on a case by case basis, but what is important in this commitment is that it shows signs of further attempts at providing an integrated and comprehensive basis for environmental regulation. In other words, sustainability foreshadows further increases in the stringency of environmental control.

5   Some Important Issues

Most of the remainder of this book deals with the detail of those environmental issues in which the power sector is intimately involved. The remainder of this chapter does not attempt to give a comprehensive view of these issues, nor does it dwell in any detail on the science of environmental diagnosis or the technology of environmental control. Rather, it picks out the three power-related environmental issues of recent years that arguably have had the greatest political prominence, namely acid rain, nuclear waste disposal, and climate change. It then outlines the main features of the debates from a policy and economic rather than a scientific perspective. One purpose here is to show how these big issues involve considerations that are much wider than the purely scientific. The presence of these large political factors is partly because of the inherent importance of the issues themselves—with all their potential for economic gain and loss to different parties—and partly because the science is often so complex and difficult that clear scientific guidelines are not always available.

Acid Rain

The term 'acid rain' (well over 100 years old) has come to encompass a wide range of phenomena. Narrowly, it refers to the wet deposition of sulfur dioxide and nitrogen oxides, but the term now has a wider political meaning and is often taken to include wet or dry deposition of any group of potentially acid-forming

---

[11] Department of the Environment, *This Common Inheritance : Britain's Environmental Strategy*, HMSO, London, 1990.

[12] World Commission on Environment and Development, *Our Common Inheritance*, Oxford University Press, Oxford, 1987.

substances or precursors. However, the main regulatory focus has been on the control of SO_2 and NO_x. The range of damage attributed to acid rain is wide: it includes acidification of lakes and streams, changes in the acidity of soils and surface waters, damage to fish life, and (especially in Germany) damage to forests. Sulfur dioxide also has well-known effects on buildings and human health at high enough concentrations, but in recent years local urban concentrations have been relatively low, and attention has focused on the long-range, often transboundary, character of acid rain.

A wide range of technologies is available to reduce the emissions of both sulfur and nitrogen, including: coal cleaning, which can remove up to half of the sulfur; combustion modification, which can reduce NO_x production; flue gas desulfurization, which can remove over 90% of all sulfur that would otherwise be emitted; selective catalytic reduction of NO_x, which can reduce emissions by up to 85%; and a range of cleaner generation technologies, especially those based on natural gas (sulfur-free) plus renewable energy and, at least in principle, nuclear power.

Concern with the effects of sulfur pollution in urban areas began in the inter-war period. In 1930, the UK became the first country in the world to operate FGD systems on power stations. Three London power stations were fitted with FGD between 1930 and 1963 in order to protect the parks and buildings of central London from the effects of acidity in emissions. However, the systems installed were neither very effective nor cheap (especially at Battersea) and the practice was not diffused more widely. When the Clean Air Act was passed in 1956, there was no requirement on generators to fit FGD or any other specifically sulfur-reducing technology. At this time there was no concern that sulfur-based or other acid deposition would reach beyond national boundaries. As power stations became located away from rural areas and were built with very tall stacks, earlier urban problems from electricity generation abated.

Modern concern about acid rain in Europe dates from 1972, when the UN Conference on the Human Environment in Stockholm was presented with Swedish evidence about damage to soils and lakes as a result of long-distance transport of pollutants. There was initially a great deal of scepticism in the international scientific community about this evidence, but intensive study over the next 15 years led to a major reversal in opinion, and acceptance that power station emissions of acid precursors were a significant cause of damage over long distances, though the precise mechanisms are considerably more complex than was first thought.

The last country to accept the evidence of the damage caused by sulfur and nitrogen was the UK. The experience of the early scrubbers in London, plus evidence of poor FGD performance elsewhere in the 1970s, convinced the CEGB and others in the UK that FGD was a dead-end. Improvements in emissions would need to await the commercialization of advanced technologies like fluidized bed combustion, which would simultaneously improve both environmental performance and operating efficiency. There was also little need for new plant, and the cost of retrofitting FGD was much higher than its incorporation into new plant. When negotiations on the LCPD started in Europe after 1984, the initial UK stance was that there was no good evidence to link UK or other emissions with long-range damage to forests or lakes. Germany was the Member State most interested to instigate tighter Europe-wide standards: German industry was keen to have a 'level playing field' of similar environmental standards throughout the Community. Concerned about the possible impact of emissions on 'forest death', Germany had passed stringent rules domestically in 1984 for the control of acid precursors and by 1988 some 36 000 MW of coal- and lignite-fired plant had FGD fitted.

As it became evident that Britain alone was holding out against agreement on the LCPD, the CEGB unilaterally offered a programme of 6000 MW of FGD, and finally in 1988 an agreement was reached. The UK emission requirement—reductions on the 1980 level of SO_2 of 20%, 40%, and 60% by 1993, 1998, and 2003, respectively—was less stringent than for the other 'Northern' Member States (whose reductions were to be 40%, 60%, and 70%, respectively), and without such a concession it was doubtful whether the UK would have signed up. The understanding was that these emission reductions would necessitate a total of 12 000 MW of FGD, though in practice only the CEGB's original 6000 MW have so far been installed. NO_x was also included in the LCPD but was a much less contentious issue.

Although sulfur emissions derive from a number of sources, the power sector is responsible for over 75% of all sulfur emissions from stationary sources in the UK. As power stations are easily the most controllable source of emissions it was inevitable, perhaps, that the main generators were to bear the full weight of the reductions, and the new companies National Power and PowerGen were given 'bubbles'—maximum total values of emissions for future years that would allow the UK to meet its requirements under the LCPD.

In the 1990s, political interest in the acid rain issue has declined sharply. This is partly because, after the long-delayed signing of the LCPD, the issue seemed largely closed. In addition, the issue of global warming and climate change caught the imagination of politicians and public after the famous speech that Mrs Thatcher gave to the Royal Society late in 1988, drawing attention to the seriousness of climate change and proposing that the UK should take an international lead. Attention in acid rain now focuses on the 'how' questions rather than the 'whether' question. The UK signed up in 1994 to the Sulphur Protocol of the ECE Convention on Long-Range Transboundary Air Pollution (LRTAP), which commits the UK further to a 70% reduction in sulfur emissions from large plant by 2005 and 80% by 2010. These superficially appear to be demanding targets. However, the rapid switch in the UK power station stock away from coal-firing and towards the combined cycle gas turbine (CCGT, fuelled by sulfur-free natural gas) means that there will be no serious problem in meeting this new obligation.[13]

By comparison with the other two cases discussed subsequently in this chapter (nuclear waste and climate change), acid rain is a relatively bounded environmental problem. While the science needed to understand it was (and is) complex, it was nevertheless in the end possible to reach reasonable scientific consensus internationally, and fortunately there were a number of 'technical fixes' available

---

[13] J. Surrey (ed.), *The British Electricity Experiment*, Earthscan, London, 1996.

at relatively low cost (initially FGD, and more recently the CCGT), so that the level of 'pain' involved in finding solutions was limited. Despite these relatively advantageous framing conditions, it is interesting that acid rain still became a major political issue, in which the stance of national Governments—especially in the UK—for a long time owed much more to a perception of national economic interest than to the scientific evidence.

Nuclear Waste

The problem of nuclear waste management and disposal has been significant in all countries that have used civilian nuclear power. In the UK, it is conventional to divide nuclear waste into the three categories of

- Low Level Wastes (LLW), which require only limited amounts of special handling and have always been expected to be disposed, when in solid form, near the surface (or at sea)
- Intermediate Level Wastes (ILW), which pose greater handling problems and have been subject to a wide variety of disposal suggestions
- High Level Wastes (HLW), which are heat-generating and require extremely careful and heavily engineered handling and disposal

Compared to waste products in other energy industries, the volumes involved are quite small, and in the area of HLW, the volumes are tiny. Radiation and the hazards it poses for humans are among the most intensively researched of all potential environmental and safety dangers. Well before the self-sustaining nuclear chain reaction had been demonstrated by Fermi in 1942, there were clear indications of the hazards of radioactivity. By 1928, the ICRP had been set up as the first international body devoted to radiation protection, and in 1934 it made its first recommendations on tolerable levels of radiation exposure. By the time power reactors became commercialized in the 1960s, the body of knowledge about radiation effects was large, and has steadily grown larger still. While this section concentrates on the management of wastes under controlled conditions, one of the obvious reasons for the intense politicization of the nuclear environment issue is the immense potential for harm if there is an uncontrolled release of radioactivity from a reactor.

Partly because of the large body of scientific knowledge about nuclear radiation and its effects, many scientists, both within and outside the nuclear industry, are convinced that there is a sound technical base for dealing with nuclear wastes. Nevertheless, the science is not without controversy, especially on the issues of the long-term human health effects of low doses of radiation and the hydrogeology of deep disposal sites, and this is partly why solutions have proved so elusive.

The history of nuclear waste management policies in the UK has been dominated by their piecemeal and changeable character. The main tension in UK policy has been between an early 'dilute and disperse' philosophy and a more recent 'containment' approach, with containment now having broadly won out in recent times.[14] The only lasting element in policy has been a commitment to the

[14] F. Berkhout, *Radioactive Waste: Politics and Technology*, Routledge, London, 1991.

Drigg site (near Sellafield) for the disposal of solid LLW.

The landmark report in UK nuclear waste policy came in 1976 with another RCEP publication, the so-called 'Flowers' report,[15] named after its chairman Sir Brian (later Lord) Flowers. The report was comprehensive and authoritative and represented the first real attempt to systematize nuclear waste policy and relate it to nuclear power policy more widely. It enunciated the 'Flowers criterion' that no commitment should be made to a large programme of nuclear power until a safe method for the containment of radioactive wastes had been demonstrated.

In the immediate post-1976 period, the influence of the Flowers report was limited, and dilution/dispersion remained the main official policy. This was especially marked in the commitment to an expanding policy of sea dumping of some categories of LLW and ILW, while simultaneously searching for deep land sites for high level waste. A major test drilling programme to look for suitable deep repository sites had to be abandoned in 1982 because overwhelming public pressure made it impossible to continue. At the same time, decisions in favour of vitrifying HLW were accompanied by advice to Government from RWMAC to wait for up to 50 years before seriously considering disposal options for HLW.

A series of increased liquid discharges at Sellafield in the early 1980s led to Department of the Environment pressure on BNFL to *reduce* discharges, rather than simply stay within limits. This marked a move toward containment and away from dilution as a policy response. It led to a large new investment programme to reduce emissions at Sellafield that lasted well into the 1990s. At one level this could be seen as simply a tightening of the ALARA principle that guided radioactive waste management policy, but in reality it was also a highly political decision, resulting from external pressure (*e.g.* the Irish Government) as well as internal dissent. Together with the political need to abandon sea dumping of nuclear waste at around the same time, the 'containment' approach was beginning to predominate over dilute and disperse.

After the abandonment of deep drilling for repository sites, the industry-funded NIREX began a new search for near-surface repositories for both LLW and ILW. This search provoked considerable hostility near the sites, and a 1986 Environment Select Committee Report—highly critical of official policy—recommended that near-surface stores be confined to LLW. It also recommended that a 'Rolls Royce' approach be taken to nuclear waste issues to overcome public suspicions.[16] As a result of this report, NIREX did abandon the idea of using near-surface repositories for ILW, but the 1987 General Election campaign overtook the whole programme, which was entirely abandoned only weeks short of the Election date.

After the 1987 Election, policy changed yet again, and now NIREX embarked on a new search for *deep* repository sites, again for LLW and ILW. Within a short period only two sites emerged as serious candidates, Sellafield and Dounreay. It was evident—given earlier fierce opposition by other local communities—that these sites had been chosen almost entirely because of their presumed acceptability

---

[15] Royal Commission on Environmental Pollution, *Nuclear Power and the Environment*, 6th Report (Flowers), HMSO, London, 1976.
[16] House of Commons Environment Committee, *Radioactive Waste*, First Report, Session 1985–86, HC-1191, HMSO, London, 1986.

to public opinion, and hardly at all on grounds of geological acceptability. Dounreay and Sellafield were two remote communities that depended entirely on the nuclear industry for local livelihoods. The search then narrowed to just one site at Sellafield (where over 60% of all waste for deep disposal already resided at the surface), and the LLW element of the proposal was abandoned. This abandonment was mainly due to the fact that the existing LLW site nearby at Drigg could now, with the use of supercompaction, continue to accept waste much further into the future than had previously been assumed.

NIREX applied for planning permission to build a so-called Rock Characterization Facility (RCF) as a large underground laboratory at the site, and a Public Inquiry was held on the proposal in 1995/96. The Inquiry showed how deep the scientific divisions ran. These divisions were here not about radioactivity and its health impact but mostly about hydrogeological predictions covering thousands of years into the future. The main issue was the possibility of predicting the probabilities of radioactivity leaking from the repository and finding its way back to the human environment over very long periods. NIREX assembled a considerable weight of scientific expertise, but the opposition (especially Friends of the Earth) also produced a highly credible array of mostly academic witnesses on relevant scientific areas.[17] The judgement from the Inquiry Inspector was severe on NIREX. On several grounds, partly scientific and partly to do with other inadequacies in the presentation of their case, NIREX were refused planning permission by the Inspector, and the Secretary of State immediately endorsed these findings. Once again, nuclear waste policy was in disarray and it is still not at all clear, despite a major White Paper on radioactive waste management policy as recently as 1995,[18] what direction future policy will take. At present the only settled disposal route for radioactive waste in the UK is the solid LLW site at Drigg—everything else is again in the melting pot.

Nuclear waste in many ways has excited even stronger feelings than acid rain had done, and as in other countries, nuclear waste policy has become highly politicized. This politicization runs partly along the fault line between nuclear supporters and the nuclear opposition. However, it has also been about the strong hostility of local communities close to proposed sites, itself a reflection of a wider lack of trust on the part of significant citizen opinion about the scientific and other information produced by the nuclear industry. There has also been significant scientific disagreement across several disciplines to throw into the balance as well. While some argue that there is in principle just as good a 'technical fix' to the environmental problems of nuclear waste management as for acid rain, the complex and often highly charged politics of nuclear waste has prevented the UK from even starting to find acceptable environmental solutions to the greater part of the problem.[19]

---

[17] R. S. Haszeldine and D. K. Smythe (eds.), *Radioactive Waste Disposal at Sellafield, UK*, Department of Geology and Applied Geology, University of Glasgow, 1996.

[18] Department of the Environment, *Review of Radioactive Waste Management Policy: Final Conclusions*, Cm 2919, HMSO, London, 1995.

[19] M. Sadnicki and G. MacKerron, *Managing UK Nuclear Liabilities*, STEEP Special Report No. 7, Science Policy Research Unit, University of Sussex, 1997.

Climate Change

Within the electricity generating industry, the largest single problem now faced is undoubtedly climate change and the possible actions that might be taken to prevent its worst potential impacts. The merits of scientific predictions decades ahead about rates of change of temperature, and potentially new and less stable climatic patterns, remain a subject of legitimate scientific debate and considerable disagreement. However, the political reality after the Kyoto agreement of December 1997 is that the industrialized countries are now likely to accept legally binding commitments to reduce carbon greenhouse gas emissions (GHGs) to around 5% less than the 1990 levels by the period 2008–2012. This represents a radical departure for the world community, and while the power sector is only one of a large number of contributors to the production of GHGs, it seems certain that the generating industry will be expected to play a very large role in emission reductions.

The science and politics of climate change are vast subjects and changing rapidly. Only a very short treatment can be offered here. As in the case of acid rain, it has been observed since the 19th century that increasing carbon dioxide concentrations as a result of expanding economic activity would lead, all else equal, to rising global temperatures because less solar radiation is reflected back into space.[20] Carbon dioxide concentrations in the atmosphere appear to have risen from some 280 ppm in the 18th century (*i.e.* before significant industrialization) to around 350 ppm in the late 1990s, with some acceleration in recent years.

It has become apparent during the 20th century that there are, in fact, a number of GHGs, all acting in a broadly similar fashion in retaining a higher proportion of solar radiation. Carbon dioxide remains easily the most important (almost 80% of the UK's warming contribution) followed by methane (12% of UK contributions) and nitrous oxide (4%).[21] CFCs, ozone, and water vapour also play a part, together with more than 30 other gases with individually very limited effects. The proportions of each gas vary by country, and estimates of the relative importance of different GHGs also vary, but it is widely agreed that CO_2 accounts for the bulk of all worldwide contributions to warming effects. As the great bulk of carbon dioxide derives from fossil fuel burning, and the power generating sector of many countries is a dominant source of fossil fuel combustion, the role of electricity generation in policy is inevitable, especially as the other major contributor—vehicle emissions—are much harder to control, both technically and politically.

The models which predict future weather and temperature patterns are immensely complex and have great difficulty in making sensible predictions about regional weather effects. Early models have tended to predict an average level of temperature rise in the range 1.5–4.5 °C, and later work has confirmed this range, but there is still some doubt about the number of decades into the future that such effects will take place. More recently there has also been more interest in

[20] P. Brackley (ed.), *World Guide to Environmental Issues and Organisations*, Longman, London, 1990.

[21] O. Greene and J. Skea (eds.), *After Kyoto: Making Climate Policy Work*, Special Briefing No. 1, Economic and Social Research Council Global Environmental Change Programme, Science Policy Research Unit, University of Sussex, November 1997.

the possible greater instabilities and extremes in future climate patterns, factors which may have large disruptive effects of their own. Relatively few scientists (apart from some in the pay of some commercial interests, especially in the USA) now seriously dispute the directions of *change* for future climate, and the UK Government Chief Scientist has recently published strong confirmation of this.[22]

International policy activity on climate change is of recent vintage. The United Nations Environment Programme (UNEP) was the first to become involved, with a series of international conferences from 1985 onwards. In 1988, UNEP and the World Meteorological Organization (WMO) jointly sponsored the setting up of the Intergovernmental Panel on Climate Change (IPCC), which has focused on three main areas through working parties:

- The scientific evidence on climate change
- Impacts of climate change on the environment and agriculture
- Response strategies

Mrs Thatcher's speech to the Royal Society in 1988 gave climate change a strong political impetus in the UK. Attention soon turned to the first 'Earth Summit' at Rio in 1992, which developed the Framework Convention on Climate Change (FCCC). Industrialized countries made 'soft' (non-legally binding) undertakings to stabilize their year 2000 emission levels at 1990 levels by the year 2000. The great majority of these countries will fail to achieve this target, generally by around 8–10%. However, the UK—largely through the substitution of gas for coal in the power system—will almost certainly achieve a largely fortuitous reduction in emissions by 2000 of over 5% compared to 1990 levels.

The Kyoto meeting in December was the third 'Conference of the Parties' to the FCCC, and was the landmark event in international climate change policy. At Kyoto the industrialized countries (Annex 1 countries) began to commit themselves to emission limits for the 2008–2012 'commitment' period that will be legally binding. For the first time, the commitment is to a *reduction*, in a basket of emissions from six gases, by roughly 5% for the commitment period. Within this global effort, there are significant variations: 8% is the 'normal' reduction level, but the USA (7%), Japan, and Canada (6%) have smaller commitments, and Russia, Ukraine, and New Zealand need only stabilize at 1990 levels.[23] As yet, the developing countries have no commitments. The international mechanisms for achieving these reductions are far from clear and await future conferences (the first important one at Buenos Aires in November 1998), but there are outline proposals for several different kinds of emissions' trading mechanisms to allow for flexibility in the achievement of the global commitments. The new Labour Government announced its commitment to a 20% fall in carbon dioxide emissions by 2010 and, although this appears now to be turning into a 'target' rather than a 'commitment', the UK is still keen to be seen to be taking a leading position in the international arena.

Climate change policies are much more complex than for any other environmental

---

[22] R. May, *Climate Change: A Note by the UK Chief Scientific Adviser*, Office of Science and Technology, Department of Trade and Industry, London, 1997.

[23] ENDS, *The Unfinished Climate Business after Kyoto*, Report No. 275, London, 1997.

issue. There are two main reasons for this. First, the problem is truly global, so that actions in any one part of the world will have very little effect unless co-ordinated with actions more widely. This means that for climate change policy to be effective the developing countries will in some way need to be incorporated into a control regime, and it makes the complexity of policy making unprecedently large. Second, there is no clear technical fix for climate change—nothing resembling FGD plants for acid rain or deep repositories for nuclear waste. Considerable efforts are being expended on carbon sequestration technologies (*e.g.* injection of carbon into deep oceans or dry gas reservoirs) and these could in time make a contribution, but there is no real substitute for finding ways of emitting fewer gases over long periods of time. This means reducing fossil fuel use, and this poses immense challenges in a world which would, left to itself, consume ever larger amounts of fuel sources that are now reckoned more plentiful than 20 years ago.

In the UK, there has been a temporary respite from some of these problems, resulting from the substitution of natural gas in CCGTs for coal firing in power generation. The higher efficiencies of the CCGT and the lower carbon content of natural gas than coal mean that this substitution leads to more than halving carbon emissions per unit of electricity generated. While some further substitution of gas for coal is possible, the further scope for this so-far painless option is limited. Clearly the power sector is a prime target for climate policy: on the supply side, there is also the prospect of developing more or less carbon-free technologies to replace fossil combustion. This clearly includes the currently much-favoured renewable technologies, and could in time include the less-favoured nuclear option. Within the power sector, there is also the prospect of better efficiency in energy use, where the prospects seem large but realization more difficult.

Most difficult of all is the transport sector, currently responsible for 25% of UK carbon emissions, and which alone shows consistent emissions growth. Government expectations are that emissions will grow between 1994 and 2020 by 23% to 36% in the absence of new policy action. The technical difficulties here are large—popular solutions like public transport improvement are expensive and make modest contribution to the overall problem—and the political difficulties of attempting to restrict private motoring make this a particularly tough problem. In practice this will almost certainly mean, as with acid rain, that the power sector will bear a share of burdens disproportionate to its emission contribution.

Climate change illustrates the political forces at work in environmental policy in the acutest form. There are two powerful levels at which this works. First, the need for international and even global agreement means that inter-Government negotiations are the essence of climate change policy, and while these negotiations are influenced by science, the nature of the deals done at Rio and Kyoto were fundamentally political. Politics are also in the lead at the level of domestic policy implementation because of the virtual absence of technical fixes. This means that action to control GHG emissions are potentially difficult, expensive, and (especially in the transport sector) highly controversial. Despite this predominance of political forces, climate change actions also represent something of a triumph for science. There is no previous history of increasingly serious, globally co-ordinated action to solve a 'problem' that really only exists in the predictions by scientists of *future* problems.

6   Conclusions

As environmental concerns at different levels have grown continuously since the 1920s, so the electric power sector has been implicated in many of the most important issues. Starting with the rural visual intrusion of overhead lines in the 1920s and 1930s, and moving through acidification and nuclear waste to climate change, the contribution of electricity generation and transmission to the problems has been large.

For some of the earlier and more local issues, the cause and the solution of the problems were simultaneously with the electricity industry. Much the same is true of the nuclear waste management problem which, while politically very important, is also a technically relatively confined issue. However, as environmental issues have broadened into continental and global dimensions with acid rain and climate change, so the contributions of electricity, though still large, have become one among several sources.

Given that the political forces in the solution of environmental problems become more and more prominent as the scale of the problem enlarges, it has also become evident since the acid rain issue that the electric power sector would be called on to play a role in the solution of problems that is disproportionately large. This is largely because of the high level of 'controllability' of the electricity supply industry, itself a reflection of the large unit sizes of plant, and a tradition of public ownership and later public control of the industry. Whether from a wider social perspective this is good or bad is moot: it certainly guarantees that electricity will remain in the cockpit of environmental policy as the climate change issue is likely to start to inflicting economic pain at the start of the new millennium.

Impact of Power Generation on Air Quality

BERNARD E. A. FISHER

1 Introduction

LONDON Implacable November weather. Smoke lowering down from chimney-pots, making a soft black drizzle, with flakes of soot in it as big as full-grown snow-flakes—gone into mourning, one might imagine, for the death of the sun. Fog everywhere. Fog up the river, where it flows among green aits and meadows; fog down the river, where it rolls defiled among the tiers of shipping, and the waterside pollutions of a great (and dirty) city. Fog on the Essex marshes, fog on the Kentish heights. Fog in the eyes and throats of ancient Greenwich pensioners, wheezing by the firesides of their wards . . .

(from the first page of *Bleak House*, Charles Dickens, 1852)

Air pollution was long regarded as a traditional feature of London life. Measures to improve air quality in urban areas have inevitably been focused on what are seen as the largest visible emitters, the tallest chimneys. Power station operations have always been associated with large scale operations in terms of fuel burnt and quantities of pollutant emitted. The power industry has thus needed to justify very carefully its impact on air pollution because of the scale of its operations.

The development of air pollution control in Britain was strongly influenced by the occurrence of a London smog on 5–8 December 1952. This consisted of stationary fog over the city into which smoke and sulfur dioxide (SO_2) from domestic chimneys mixed. Subsequent analysis showed that over 4000 extra deaths had occurred as a result of the air pollution episode. Figure 1 shows an example of the analysis which confirmed a relationship between air quality and health.

Large numbers had died in earlier London smogs of 1873, 1880, 1881, 1882, 1891, and 1901.[1] What distinguished the 1952 smog was the public attention it generated. A committee led by Sir Hugh Beaver was set up. Their recommendations formed the basis of the parliamentary Clean Air Act in 1956, whose objective was

---

[1] D. Elsom, *Atmospheric Pollution*, Blackwell, Oxford, 1987.

Issues in Environmental Science and Technology, No. 11
Environmental Impact of Power Generation

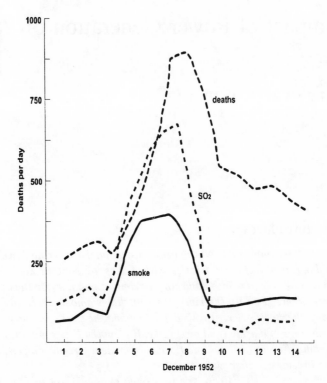

Figure 1 Daily air pollution levels and deaths in London, December 1952

to restrict emissions from domestic fires. Smoke control was made a local option and not a statutory duty. Local authorities were encouraged to designate 'smoke control areas' in which only authorized smokeless fuels could be burnt. The cost of converting heating appliances to burn smokeless fuels was shared between the government, the local authority, and the householder.

In 1952 there were 29 operational sites in London involved with public power generation, consisting of 38 coal-fired power stations with a maximum generating capacity of 2500 MW and an average thermal efficiency of 22%. The typical stack height was 300 feet. The new clean air legislation covered industrial chimneys. However, the larger, more difficult industrial processes such as power stations remained under the central control of the Alkali Inspectorate. There was thus a distinction between central and local control of air pollution. Power stations were not considered to have had a major role in the 1952 London smog episode. Plumes from most power station stacks would have risen above the shallow mixing layer in which the urban pollution was trapped.[2]

At the same time as smoke control areas were introduced in city centres, the growth of the power industry led to the development of larger, more efficient power stations, which were located outside of urban areas. This was accompanied by more systematic investigations of the dispersal in the atmosphere of the air pollutants from tall chimneys. At first, systematic surveys of dust, or particle deposition, were conducted around major rural power stations, such as those at

[2] G. Spurr, *J. Meteorol.*, 1959, **16**, 30.

Little Barford[3] and Staythorpe.[4] The effectiveness of measures to control the emission of dust through the installation of electrostatic precipitators was demonstrated. Systematic surveys of SO_2 around power stations followed and continue to this day.[5]

Reviews of progress in smoke control were made by the Royal Commission on Environmental Pollution in 1974 and 1976. Improvements in air quality could be demonstrated from the results of the so-called National Survey of Air Pollution which consisted of some 1200 monitoring sites for smoke and SO_2 distributed throughout the UK with most in urban areas. Smoke is defined as 'particles' resulting from incomplete combustion or chemical reaction, and which are usually less than several microns in diameter. The method of measurement actually determined soiling, or blackness, in the case of smoke, and net acidity in the atmosphere in the case of sulfur dioxide. Between 1958 and 1972, average concentrations in urban areas had decreased dramatically[6] with an almost parallel drop in emissions and concentrations (see Figure 2). The industrial requirements of the Clean Air Act had been largely met by 1960 and were followed by the substitution of coal by oil after 1960. The domestic requirements of the Act were supplemented by the switch in domestic fuel use to convenient fuels, like electricity, oil, and especially gas. Decreasing differentials in the price of fuels had also played a part.

The power industry was more concerned with the concentrations of the acid gas SO_2 around power stations, which is released from power stations in much greater quantities than dust. None of the measures arising from the Clean Air Act was specifically directed to SO_2, which is emitted as an inevitable consequence of burning coal and oil which usually contain a small fraction of sulfur (typically 1–2% by weight). However, the changes in domestic fuel usage, the closure of smaller urban power stations, and the building of larger power stations with tall multiflue chimneys, with a typical height of 200 m led to dramatic decreases in urban SO_2 concentrations.

Role of Local Authorities

When the Beaver Committee produced their report, they published a map showing areas they considered to be 'black', where the need for improvement in smoke was most urgent. The Government drew up a list of local authorities whose areas were wholly or partly 'black' on the basis of the Beaver map. In England there were about 300 such areas, and by the time of local government re-organization in 1974 the majority of the local authorities concerned had made some smoke control orders, with many well on the way to total coverage. The distinction between 'black' and 'white' areas was administrative: the whole of an

---

[3] G. England, C. J. Crawshaw and H. J. Fortune, *J. Inst. Fuel*, 1957, **30**, 435.

[4] P. J. Meade and F. Pasqill, *A study of the Average Distribution of Pollution around Staythorpe*, Meteorological Research Committee Paper Number 1039, Meteorological Office, Bracknell, 1957.

[5] D. Laxen, *Generating Emissions? Studies of the Local Impact of Gaseous Power-station Emissions*, National Power, 1996.

[6] Royal Commission on Environmental Pollution, *Fourth Report Pollution Control: Progress and Problems*, Cmnd 5780, HMSO, London, 1974.

old borough was normally categorized as black or white and the list of black areas was never updated. When local government was re-organized into larger areas, the list became valueless: new authorities included old 'black' areas as well as large rural areas where smoke control would be pointless. Local authorities were left to decide where smoke control was needed without central guidance.

A disincentive to smoke control areas raised in the Royal Commission's 1976 report was the notion that smoke drifts so far downwind that a single authority in the middle of a large polluted region would be wasting its time if it imposed smoke control. The Commission considered whether sufficient guidance was available to local authorities in deciding whether smoke control should be introduced.[7] It recommended that the Government should draw up guidelines to assist in the decision making process. This should take account of factors such as the amount of smoky fuel burnt in an area, population density, topography, and climate. No such systematic guidance was produced, probably because it was not seen as a priority. The Commission also recommended that the decision on smoke control areas should depend on measurement and an assessment of the factors involved.

By 1982 there were six small coal- and oil-fired power stations left near the centre of London and all closed in the next few years to be replaced by modern coal- and oil-fired 2000 MW power stations with an efficiency of 36%. Combustion gases are discharged through a single stack of height 200 m. For these power stations dispersion was recognized to occur over distances of some tens of kilometres. The maximum ground-level concentration would occur at a distance of about 10 km from the stack. A high degree of spreading and dilution of emissions before the plume touched the ground was achieved.

The Weather and Smog

The benefits of smoke control on human health has been demonstrated by the reduction in the number of extra deaths brought about by episodes of adverse weather conditions, although the actual causative factor is not known. Table 1 illustrates the conditions in London on three occasions, each in December, when pollution was unusually high. The concentrations of smoke and SO_2 are expressed in units of $\mu g\,m^{-3}$. Lawther and Bonnell[8] caution against ascribing the reduced effect in 1962 wholly to the reduction in smoke. In 1962, patients were publicly discouraged from venturing outdoors and methods of treatment of respiratory and cardiac distress had improved.

The London fog of 2–5 December, 1957, is another example of adverse weather influencing pollution levels.[9] At Woolwich, concentrations of smoke reached 5430 and 5220 $\mu g\,m^{-3}$ on 4 and 5 December, while sulfur dioxide concentrations were about 2700 $\mu g\,m^{-3}$ on both days. These levels are comparable with the maximum concentrations recorded in central London in the 1952 episode.

[7] Royal Commission on Environmental Pollution, *Fifth Report Air Pollution Control: An Integrated Approach*, Cmnd 6371, HMSO, London, 1976.

[8] P. J. Lawther and J. A. Bonnell, *Some Recent Trends in Pollution and Health in London and Some Current Thoughts*, Proceedings of the International Air Pollution Conference of the International Union of Air Pollution Prevention Associations, 1970.

[9] A. E. Martin, *Int. J. Air Pollut.*, 1959, **2**, 84.

Table 1 Comparison of three December air pollution episodes in London

Year	Maximum mean daily concentrations in Central London/$\mu g\,m^{-3}$		Estimated extra deaths in Greater London
	Smoke	SO_2	
1952	> 6000	3500	4000
1962	3000	3500	750
1972	200	1200	not detectable

Deaths due to the fog in Greater London were estimated to be between 760 and 1000. This excluded the 87 deaths in the Lewisham train disaster on 4 December, 1957, an indirect result of the fog!

Meade[10] comments that it needs only a suitable weather situation in winter, associated with the large central area of an anticyclone, very light winds or calms, and a severe subsidence inversion aloft for there to be a serious risk of smog. Winter is the important season because the heat from the sun may not be strong enough to disperse the fog and clearance has to wait the arrival of freshening winds when the anticyclone moves away. The constant factor is the smoke itself, which is discharged to the atmosphere throughout the winter from chimneys all over the country. The variable factor is the weather. From year to year there are variations in the positions of anticyclones. The weather situation in the London smog of December 1952 was unusual in its persistence and one cannot predict when such situations may occur again.

Accepting man's inability to control the weather, Meade argued that for any district in Great Britain there are a few days each year when smoke may be trapped for many hours in a shallow layer and cause smog. On these days, great benefit would be gained if little or no smoke was emitted. In contrast, on many days it would not matter whether domestic or other chimneys emitted smoke. This meteorological control of emissions as an intermediate step pending the widespread adoption of smoke control areas was never taken up because of the practical difficulties. In the weather conditions associated with smog episodes, power stations with tall stacks located outside urban areas were not thought to be implicated directly.

2 Trends in Emissions

The objective of the Beaver Committee's recommendations was that by the end of 10–15 years the total smoke in heavily populated areas would be reduced by about 80%. The Royal Commission in 1976 did not feel that it was able to quantify to what extent that target had been met after 20 years. However, the trends illustrated by Figure 2 show that a great deal had been achieved and certainly after 1976 smoke was not considered to be a major air pollution issue.

Smoke control had the effect of reducing the emissions of sulfur dioxide as well as smoke, although it was not a stated aim of the Clean Air Act. Sulfur dioxide emissions from domestic sources decreased by 50% between 1950 and 1972 as a

[10] P. J. Meade, *Int. J. Air Pollut.*, 1959, **2**, 87.

Figure 2 Smoke emissions and urban concentrations in the UK, 1958–1972

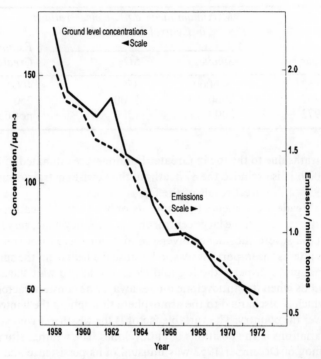

consequence of the changes in domestic fuel use. The Royal Commission in 1976 recommended a policy of only authorizing fuels with a low sulfur content in smokeless areas to reduce, as far as practicable, low-level emissions of sulfur dioxide.[7] From the point of view of ground-level concentrations of sulfur dioxide in urban areas, it was thought preferable for high sulfur fuels to be burnt in large industrial furnaces, such as power stations, where combustion is efficient and pollution could be dispersed through tall chimneys so that the resulting ground-level concentrations would be low. National emissions of sulfur dioxide (SO_2) peaked between 1965 and 1970 at 6.5 million tonnes SO_2. The policy was carried through by the completion and operation of a number of large oil- and coal-fired major power stations, whose emissions peaked in 1979 at 3.2 million tonnes SO_2.[11]

3 Dispersion and Dilution

For local authorities, advice on the determination of chimney heights was provided by the Memorandum on Chimney Heights,[12] the third edition of which was produced in 1981. The Alkali Inspectorate used a variety of similar methods for assessing major industrial processes. Major industries, such as the Central Electricity Generating Board (CEGB), adopted these methods and started to

[11] A. G. Salway, H. S. Eggleston, J. W. L. Goodwin and T. P. Murrells, *UK Emissions of Air Pollutants*, *National Atmospheric Emissions Inventory*, Technology Report AEA/RAMP/20090001/R/003-ISSUE 1, AEA, 1996.

[12] Department of the Environment, *Chimney Heights; Third Edition of the 1956 Clean Air Act Memorandum*, HMSO, London, 1981.

develop their own. It is probably fair to say that all these approaches are shrouded in some mystery. Hall and Kukadia,[13] in their report describing the background to the new HMIP guidelines on discharge stack heights for polluting emissions, shed some light on the Chimney Height Memorandum. It was based on the assumption that the chimney should be tall enough to avoid short-term acute pollution effects. The pollution of greatest concern was sulfur dioxide, so calculations were based on an assumed limit value for this pollutant of $450 \, \mu g \, m^{-3}$ for exposure times of 3–5 minutes. Account was taken of the effect of different background concentrations (defined in terms of the kind of district surrounding the installation). A simplified approach was adopted in which it was possible to express the recommended chimney height as a function solely of the sulfur dioxide discharge rate, with later refinements to take into account a wider range of conditions.

Research conducted by the CEGB provided a justification for heights adopted for the increasingly large power stations being designed in the 1960s and 1970s. It was recognized that the chimney must in any case take flue gases clear of any disturbance in the air flow due to nearby buildings, such as the boiler house, which was the objective of the $2\frac{1}{2}$ rule (chimney height at least $2\frac{1}{2}$ building height). Design heights were based on methods which involved extrapolating beyond current practice. Each phase of increasing power station size involved heights, discharge rates, volume flows, *etc.*, which had not been tested previously in the real world. Later surveys involved networks of sulfur dioxide measuring equipment distributed in arcs around power stations at distances at which maximum ground-level concentrations were thought to occur. The instruments needed to be able to respond to short-term fluctuations in concentrations as it was recognized that the direction of a plume from a chimney would fluctuate. Thus the daily recording National Survey instruments were not suitable.

It is probably true to say that the results of the research were not used in the design of new chimneys, but rather to check on the performance after a power station was built. Barrett[14] describes a statistical method based on short-term measurements which gives a complete description of a power station's effect on ground-level concentrations under all meteorological conditions. Moore[15] demonstrated a method of calculation that could be used to specify maximum ground-level concentrations for any height of source. The larger power stations already built and operating at the time would comply with a standard of $700 \, \mu g \, m^{-3}$ as an hourly average, while the smaller ones would only produce exceedences on occasional hours per year.

Power Station Surveys

Of interest are the results of the earliest reported surveys around coal-fired power stations between 1950 and 1956.[16] The aim of the surveys was to try to determine

[13] D. J. Hall and V. Kukadia, *Background to the New HMIP Guidelines on Discharge Stack Heights for Polluting Emissions*, report LR929, Warren Spring Laboratory, 1993.
[14] G. W. Barrett, *Clean Air*, 1979, **9**, 119.
[15] D. J. Moore, *Proc. Inst. Mech. Eng.*, 1975, **189**, 33.
[16] W. D. Jarvis and L. G. Austin, *Inst. Fuel Bull.*, 1957, **30**, 435.

the effect of ash particles emitted from the chimneys on dust levels in surrounding areas. It was shown from measurements using deposit gauges that the bulk of the ash particles emitted did not fall within 3 km of the stations. Further afield, the effects of the station were hardly detectable because the particles were highly dispersed within the atmosphere and mixed with airborne dust from other sources. This was the justification for building tall chimneys to ensure adequate dispersion of sulfur dioxide, although it was recommended that future surveys should be made measuring short-period average SO_2 concentrations. Dust was not considered to be a problem because of the very efficient dust arrestment equipment installed in the more modern power stations. Little carbonaceous material, detected as black smoke, would be present in the atmosphere around power stations because of efficient combustion.

For the next four decades from the 1950s onwards research and monitoring, of a very high quality, of airborne concentrations, mainly of sulfur dioxide, was undertaken. Results cannot be adequately presented in this review, but they reinforced the view that tall chimneys ensured adequate dispersion in the atmosphere, which could be regarded as having sufficient carrying capacity even in regions where power stations were grouped together.[5]

4 Fraction of Smoke and SO_2 Leaving the Country

On a wider scale, Meetham[17] had already concluded from the national network of deposit gauges that approximately two-thirds of the smoke emitted nationally in the 1940s, which would have included a large fraction from domestic sources, was blown out to sea. By the early 1970s concerns were being raised regarding the eventual fate in the atmosphere of the pollution emitted and it was the acidifying consequences of sulfur dioxide on distant ecosystems that would turn out to be of greatest concern. The key factor, at least from a British viewpoint, was what fraction of sulfur dioxide was exported from the country. Measurements of rain composition already suggested that sulfur from combustion sources was absorbed into cloud and deposited in rain. The question was: how far did it travel in dry conditions? If it could be shown that sulfur did not travel far from its source, then it could be demonstrated that the sulfur did no harm to Britain's neighbours.

Aircraft Measurements

A series of aircraft measurements were undertaken by the Meteorological Office in 1971 and 1973 to measure the flux of SO_2 out of the country. The observations consisted of average concentrations of SO_2 along a flight path off the east coast of England over the North Sea at different levels in the atmosphere roughly perpendicular to the prevailing south-westerly wind. As part of the flight plan, the vertical structure of the atmosphere was determined to establish the height to which pollution would have mixed. The flight path was chosen so as to intersect

---

[17] A. R. Meetham, *Q. J. R. Meteorol. Soc.*, 1950, **76**, 359.

Table 2 Fraction of sulfur leaving the United Kingdom

Date of flight	1 Oct 1971	22 Oct 1971	2 Nov 1971	9 Aug 1973	4 Sept 1973	7 Sept 1973
Wind speed (m s^{-1})	10.0	20.5	17.3	14.7	8.1	12.5
Depth polluted layer (m)	1050	450	600	1200	1700	1200
Measured fraction SO$_2$ leaving east coast	0.51	0.71	0.52	0.57	0.58	0.48
Fraction sulfate leaving east coast	—	—	—	0.12	0.09	0.20
Calculated fraction SO$_2$ leaving east coast	0.56	0.66	0.66	0.64	0.60	0.56

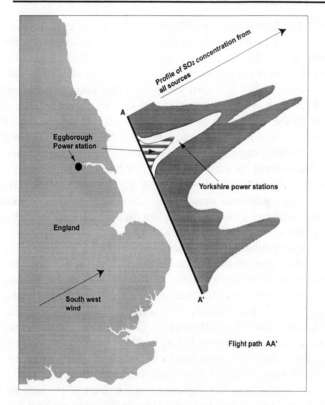

Figure 3 Typical flight path along which airborne concentrations were measured in south-westerly winds to determine the fraction of sulfur leaving the country (A and A′ are the end points of the flight path over the North Sea)

sulfur emitted from major power stations and industrial sources in the Midlands and Yorkshire (see Figure 3).

The measurements showed that the fraction of SO$_2$ leaving the country was about 0.6. A further fraction of about 0.1 of the sulfur emitted was in the form of sulfate particles (sulfur dioxide which has been oxidized to ammonium sulfate or sulfuric acid and attached to either small cloud droplets or dry particles) (see Table 2).

The value of the measurements was increased by the use of an environmental model. The contributions from a number of point and area sources with emissions estimated from the known rate of fuel combustion were added together, using the measured wind speed, wind direction, and the depth of the layer in which the pollution mixed. It was possible to calculate the concentrations along the flight path over the North Sea. The only unknown parameter is the deposition velocity, which determines the rate at which sulfur dioxide is deposited at the ground, mainly the land in eastern England. The deposition velocity is an unknown parameter in the model. By running a long-range transport model for a number of different choices of deposition velocity, it was possible to find a value which gave a good fit with the measurements. This value of $10\,mm\,s^{-1}$ is found to be consistent with small-scale measurements of the rate of deposition of SO_2 to grass and theoretical estimates of the rate of deposition to the sea surface.[18]

5   Lifetime of SO_2

With the deposition velocity known, it is possible to calculate the concentration of sulfur dioxide at various distances downwind of a major source area in dry weather conditions and so, for example, obtain an estimate of the sulfur that might cross the North Sea. This was the first step to calculating budgets of the amount of sulfur transported between the countries of Europe.

From the model it was also possible to estimate what fraction of SO_2 was leaving the country from high stacks and what fraction of the SO_2 from low chimneys was leaving the country. It turns out that although the fraction from high stacks is higher, it is not a great deal higher and certainly not a major factor in determining the amount of material crossing the North Sea. Later flights using tracer released into the stack of Eggborough power station and sensitive detection equipment were able to track the individual, overlapping plumes from power stations over the North Sea for several hundred kilometres[19] and in one case for more than 600 km.

The only other key parameter besides the deposition velocity that needed to be known to understand the atmospheric cycle of SO_2 was the rate of removal of SO_2 by rain. It was recognized that for SO_2 to be absorbed in cloud or rain droplets it had to be oxidized to sulfate, either before the onset of rain or while it was within a rain system. The early models describing the transport of sulfur oxides assumed that although the detailed chemistry was uncertain, removal by rain was efficient. Hence if SO_2 had not been removed from the atmosphere by dry deposition, it would finally be removed when it first encountered rain. The total amount of sulfur deposited over a country was the sum of dry-deposited sulfur dioxide and wet-deposited sulfur. Part of the dry and wet deposit arose from foreign sources and part from indigenous sources.

The amount of sulfur imported and exported in this way depended on a country's emissions, its size, and location. For example, over the UK, off the north-west coast of Europe, it is not hard to believe that the majority of the sulfur deposition arises from sources in the UK. This is not the same thing as saying that the majority of the emissions from the UK is deposited over the UK.

[18] B. E. A. Fisher and P. R. Maul, *The Mathematical Modelling of the Transport of SO₂ across Country*, Proceedings of the Institute of Measurement and Control Symposium: Systems and Models in Air and Water Pollution, London, 1976.

[19] B. E. A. Fisher and B. A. Callander, *Atmos. Environ.*, 1984, **18**, 1751.

6 Proportionality between Emissions and Deposition

One of the major uncertainties at the time was whether a reduction in emissions would lead to a proportional reduction in deposition of sulfur. If the process of wet deposition was controlled by the oxidation of sulfur dioxide, was it not possible that the rate of wet deposition was controlled by the concentration of oxidizing species present in the atmosphere and not by the concentration of sulfur oxides? This was the so-called 'non-proportionality' issue.[20] If the reduction in emissions of a given source by 50% would lead to a reduction in deposition from that source over a specified receptor region by between 40–60%, then the relationship between the source and receptor regions would be regarded as approximately proportional. If the reduction of emissions from the source region by 50% would lead to a reduction in deposition over the receptor region of less than 40%, then the relationship would be regarded as non-proportional and the perceived benefits of an emission control strategy would need to be carefully considered.

It was not possible to determine from deposition measurements alone whether or not a change in emission strength of a particular source would produce a proportional change in that source's contribution to deposition. However, given the reduction in national emissions of sulfur dioxide between 1970 and 1995 of 45%, with further reductions anticipated, together with the UK's isolated position from emissions in mainland Europe, it may be possible soon to determine, from trends in the measurements at UK monitoring sites, at what distances from the major coal-fired power station sources the relationship is proportional. The latest report of the UK Review Group on Acid Rain[21] does not shed light directly on this issue, though there is clear evidence to demonstrate downward trends in acid deposition. Any 'non-proportionality' is expected to be apparent in concentrations of aerosol sulfate and sulfate in precipitation. The report states that both 'have decreased at a rate similar to or slightly less than the rate of decrease in emissions'. The report shows that models based on assuming proportionality between sulfur emissions and depositions are able to broadly explain current levels of deposition. Reductions in SO_2 over the past few decades appear to be larger than the reduction in national SO_2 emissions, suggesting that the great reduction in small local sources has been of most significance.

In the absence of any appropriate experimental measurement programmes, models have been used to shed light on the non-proportionality question. Man-made emissions of SO_2 from north-west Europe in the early 1980s were about 39 million tonnes per annum, compared with 23 million tonnes from an area of north-east North America of comparable size. A simple model describing the long-range transport of sulfur oxides was optimized to give a good description of the airborne concentrations and deposition in rain measured at monitoring sites in Europe.[22] The implicit assumption in the model was that the

[20] UK Review Group on Acid Rain, *Acid Deposition in the United Kingdom 1981–1985*, Warren Spring Laboratory, 1987.

[21] UK Review Group on Acid Rain, *Acid Deposition in the United Kingdom 1992–1994*, AEA Technology Report, 1997.

[22] B.E.A. Fisher and P.A. Clark, in *Air Pollution Modeling and its Application IV*, ed. C. De Wispelaere, Plenum, New York, 1985, p. 471.

Figure 4 Total sulfur deposition over Europe $(g\,S\,m^{-2}\,a^{-1})$ in the late 1970s

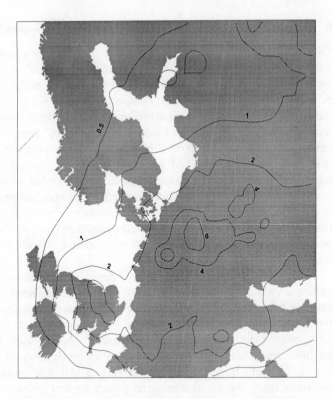

relationship between source and receptor regions was 'proportional'. The same model, with the same choice of parameters, but with allowance for different emissions and different meteorology, was then run for north-east North America. A similar degree of agreement with measurements was obtained. From this it was concluded that the sulfur system as a whole was approximately proportional, and if the sulfur emissions in Europe were reduced by approximately one half, proportional reductions in sulfur deposition would be obtained (see Figures 4 and 5).

7 Critical Loads

A critical loads map shows the sensitivity of ecosystems in Britain and Europe to sulfur deposition.[23] Emission strategies should be designed to ensure that sulfur deposition in various parts of Europe does not exceed levels which could cause harm to soils or surface waters. One consequence of the development of the critical loads concept has been that far more attention has been paid to the effects of deposition of sulfur over the UK from sources in the UK.

Estimates of the sensitivity of soils suggests that there are extensive areas in central Wales, south-west and northern Scotland, and the Pennines which are very sensitive. Sulfur deposition over these areas would have to be reduced to $0.3\,g\,S\,m^{-2}\,a^{-1}$. If those working on acid rain in the early 1970s had realized that

[23] R. W. Battarbee (ed.), *Proceedings of the Conference Acid Rain and its Impact: the Critical Loads Debate*, Ensis, London, 1995.

Figure 5 Total sulfur deposition over North America $(g\,S\,m^{-2}\,a^{-1})$ in the late 1970s

the target deposition for parts of the British Isles was as low as this, this would have upset existing notions. Considerably more attention would have been paid to the deposition from major emitters within the country and the carrying capacity of soil. For example, Figure 6 shows the calculated deposition from major coal- and oil-fired power stations when the building programme of the mid 1970s was complete. (The CEGB emissions would have been about 2.9 million tonnes SO_2.) In the event, economic and political factors led to no further increases in SO_2 emissions and the UK is currently in a period of sharp decline in sulfur dioxide emissions. Principal factors have been industrial restructuring followed by the switch to burning gas in more recent years.

8 International Agreements on Transboundary Pollution

Tackling acid rain, or more precisely acid deposition, clearly required international co-operation. Work under the UN Economic Commission for Europe started in the late 1970s and considers all major pollutants for which there is significant exchange between countries of Europe. There are three United Nations Economic Commission for Europe (UNECE) Protocols which have effect in European countries. The Geneva Protocol of 1991 on volatile organic compounds requires a reduction of the 1988 emissions by 30% by 1999. The 1988 Sofia Protocol on nitrogen oxides required NO_x emissions to remain steady. The Second Sulfur Protocol requires emission reductions in European countries to be based on the concept of critical loads. Control strategies should be designed to protect sensitive areas, so that the emissions may vary from country to country according to the country's position relative to sensitive areas. This implies that

Figure 6 Estimated deposition of sulfur ($g\,S\,m^{-2}\,a^{-1}$) from CEGB power stations from emissions when the mid-1970s construction programme of major coal- and oil-fired power stations was complete

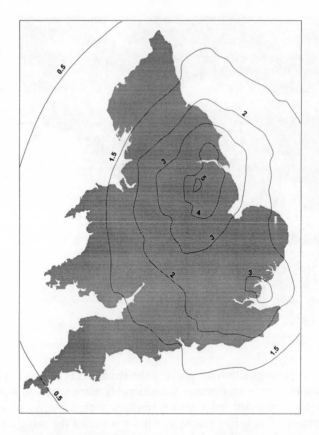

countries in northern Europe should be subject to stricter sulfur emission limits than countries in southern Europe with less sensitive soils. Sulfur emission limits under the Protocol were negotiated in 1994.

The Sulfur Protocol

Work on the Sulfur Protocol has provided source–receptor relationships between the emissions in each region of Europe and the deposition in each region. From this information, and information on the costs of emission abatement, it is possible to calculate the optimal strategy for achieving, as far as is practical, sulfur deposition lower than critical loads. Three models for determining optimum strategies have been developed: the RAINS model from IIASA (the International Institute for Applied Systems Analysis, Austria), the CASM model from the Stockholm Institute, and the ASAM model from Imperial College, London.

An integrated assessment ultimately leads to a cost–benefit relationship which shows the degree to which areas with low critical loads are protected as a function of the cost of sulfur emission abatement. It turns out that there is a turning point below which significant protection of critical regions is achieved for increasing expenditure, but above which it becomes increasingly expensive to achieve further improvement. The improvement based on a policy in which reductions vary between country to country and region to region is found to be much more effective than a straightforward reduction in each country.

Certain practical constraints emerge from the analysis. As the earlier example showed, it is not possible to achieve compliance everywhere with critical loads. Instead, the current Second Sulfur Protocol is directed towards achieving a 60% closure of the gap between current exceedences of critical loads and complete compliance with critical loads. Secondly, there is a limit to the benefit of actions taken in western Europe, if eastern European countries do not, or are unable to, take action. For Britain the 80% reduction in sulfur dioxide by 2010 represents the conclusion of 55 years of progress in tackling sulfur dioxide.

The European Union's approach to reducing sulfur emissions has been based on the 1988 Directive, which specifically addressed emissions from large combustion plant such as power stations. The UK has been able to meet the 1988 targets comfortably through a programme of flue gas desulfurisation at major power stations and a rapidly increased use of natural gas for power generation. Recent proposals from the European Commission outline much more stringent reductions than those in the Second Sulfur Protocol. Acidification effects are largely dependent on emission strength and distance to sensitive location. They do not depend strongly on height of release. Hence a policy based on total national limits of sulfur and nitrogen oxides appears more appropriate, allowing individual countries to determine the most cost-effective way to reach the limits. Within the UK, no transparent method has been developed for deciding which way to ration power plant emissions in situations when the combined deposition from a number of power stations is predicted to lead to exceedences of critical loads. This requires a balanced approach to individual power station emission limits and the emission limits on a power company as a whole.

Nitrogen and Ozone

Future work under the Convention on Transfrontier Pollution faces even more complex scientific problems. Strategies on the changes brought about in ecosystems by the deposition of nitrogen compounds need to consider both nitrogen oxides and ammonia. Nitrogen oxides are partly associated with stationary combustion sources, such as power stations (24% from public power in 1994), but also with road transport (56% in 1994), while ammonia emissions, which are greatest in agricultural areas with the highest density of animals, are not associated with power stations and are uncertain and difficult to control.

The VOC Protocol is a preliminary step to tackling episodes of high ozone. These are the incidents of summer smog associated with high ozone, particles, and oxidant concentrations, which can have adverse effects on human health and vegetation. Action on ozone requires understanding of the complex relationships between precursors' emissions of volatile organic compounds (which are not emitted from power stations) and nitrogen oxides (which are emitted from power stations). The relationship is certainly highly non-proportional. Although summer photochemical smogs involve much more complex atmospheric chemistry than winter smogs and usually extend over a wider region, both kinds of episode are driven by similar anticyclonic weather conditions and their incidence from year to year is dependent on the occurrence of the 'wrong' kind of weather and the amount of emissions.[24] Any action should involve all categories of emissions. Power stations are involved in summer smog episodes as a result of their NO_x emissions. The relationship is complex and broadly depends on the regional emissions, not the releases from a single stack.

9 Road Transport Pollution in the 1990s

Attention to the European Convention on Transfrontier Pollution led to some neglect of air pollution and health issues in the 1970s and 1980s, especially with the local sulfur dioxide and smoke problem apparently solved. Concern over the effects of some other airborne emissions, such as lead from petrol-driven road vehicles, has come and gone while other concerns remain. Attention is still paid to carcinogens, such as benzo[a]pyrene, which is present in atmospheric smoke, formerly associated with poor coal combustion and now with vehicle engines. Benzene and 1,3-butadiene are other carcinogenic organic compounds mainly related to road transport emissions.

The growth in transport movement by car within the UK, which has increased by a factor of 10 between 1952 and 1993,[25] has drawn attention to emissions from this source (see Figure 7) and reinforces the view that an integrated approach to improving air quality needs to consider all types of sources and a range of air pollutants. Road transport is subject to progressively more stringent emission controls. The introduction of three-way catalysts on new petrol-engined cars will lead to significant reductions in national emissions of carbon monoxide, nitrogen oxides, and volatile organic compounds as new vehicles fitted with exhaust controls are introduced into the traffic fleet.

The Environmental Protection Act passed in 1990 placed tight regulation on polluting industry. The operation of the Act is open to public scrutiny and also requires an integrated view of environmental protection. The approval of new and existing industrial works requires chimney emissions to meet specified limits, which are determined by applying the principles of BATNEEC, 'best available technique not entailing excessive cost', roughly interpreted as best available technology with account taken of costs and benefits. Each power station must have an authorization which can be consulted on the public register. Although the 1990 Act introduced integrated pollution control (consideration of the effects of discharges to air, water, and land), it did not directly require the combined effects of air pollution emitted from industrial, domestic, and transport sources to be considered. This needed to await the passing of the 1995 Environment Act and the introduction of air quality management.

London Pollution Episode, December 1991

It has been known for 40 years, if not longer, what causes high pollution levels. Episodes of high pollution in cities are now regularly reported in the press and

---

[24] B. E. A. Fisher, *Atmos. Environ.*, 1988, **22**, 1977.

[25] Royal Commission on Environmental Pollution, *Eighteenth Report, Transport and the Environment*, Cmnd 2674, HMSO, London, 1994.

Figure 7 Growth in surface transport movement of people by car, 1952–1993

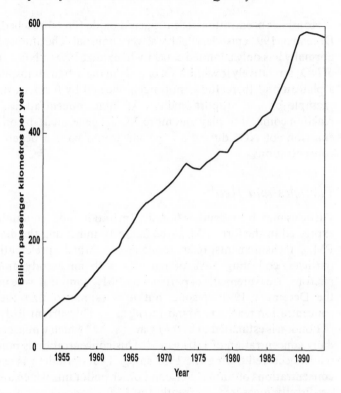

information from urban monitoring networks is readily available. It was therefore not surprising that the four-day pollution episode between 12 and 15 December 1991 attracted public interest.[26] At the time, hourly measurements of nitric oxide (NO), nitrogen dioxide (NO_2), carbon monoxide (CO), SO_2, and ozone were available. Air quality bulletins in winter refer to SO_2 and NO_2 (NO_2 is more reactive than inert NO), since ozone episodes are summertime events. CO is an indicator of traffic emissions but is not used to classify air quality, while NO and NO_2, collectively known as NO_x, arise from traffic and stationary combustion sources.

Meteorological conditions in December 1991 were very similar to earlier episodes such as that in December 1952, but the daily SO_2 concentrations levels in 1991 (about $140 \, \mu g \, m^{-3}$) were much lower, since domestic coal is no longer used in London on a wide scale. Daily black smoke concentrations, measured using the 'old' National Survey method referred to earlier, rose to about $230 \, \mu g \, m^{-3}$. The daily NO_x concentrations peaked at $1800 \, \mu g \, m^{-3}$. That NO_x and particularly NO_2 was the most significant pollutant in 1991 is not surprising. The average NO_x emission density in London is about $4 \, \mu g \, m^{-2} \, s^{-1}$.[27] In 1952 domestic coal SO_2 and smoke emissions corresponded to an estimated emission density of $10 \, \mu g \, m^{-2} \, s^{-1}$ on a cold winter day.[28]

[26] J. S. Bower, G. F. J. Broughton, J. R. Stedman and M. L. Williams, *Atmos. Environ.*, 1994, **28**, 461.

[27] M. Chell and D. Hutchinson, *London Energy Study, Energy Use and the Environment*, London Research Centre, 1993.

[28] D. H. Lucas, *J. Air Pollut.*, 1958, **1**, 71.

Hourly concentrations of nitrogen dioxide (NO_2) reached $800 \, \mu g \, m^{-3}$ in the December 1991 episode, which was very unusual. The atmosphere in urban areas normally has only a limited capacity to oxidize NO to NO_2. If the concentration of NO_x is relatively low, the level of NO_2 in the urban atmosphere quickly reaches a plateau and thereafter is not much affected by further emissions of NO_x (an example of non-proportionality). At high concentrations, another chemical reaction comes into play and more NO_2 is generated. In consequence, reducing emissions of NO_x during an episode would have a benefit in reducing NO_2 concentrations.

Particles and Health

Particles may be in solid or liquid form and in modern terminology their size is expressed in the form PM_{10} (the figure 10 indicating the diameter in microns). PM_{10} measurements refer to collectors which preferentially collect small particles, collecting 50% of particles with an aerodynamic diameter of 10 microns. Measurements of particles by PM_{10} samplers were not available during the December 1991 episode, but it is estimated that the 24-hour average concentration reached around $150 \, \mu g \, m^{-3}$. The current PM_{10} emission density of London is estimated to be $0.4 \, \mu g \, m^{-2} \, s^{-1}$,[29] which would correspond to a peak daily concentration of $180 \, \mu g \, m^{-3}$. This concentration is much lower than that which occurred during the 1952 smog. That episode was associated with high concentrations of smoke, SO_2, and other pollutants which at the time could not be directly measured. Currently the PM_{10} concentration is thought to be the component of the air pollution which is most closely associated with health effects. In the late 1970s the effects of fine particulate associated with sulfate came to attention though the USEPA's Community Health and Environmental Surveillance System CHESS, though the results of this study are discounted by the Advisory Group on the Medical Aspects of Air Pollution Episodes.[30].

The number of excess deaths from respiratory and associated symptoms in London in the December 1991 episode was estimated to be between 100 and 180, equating to an increased death rate of around 10%. The normal incidence of deaths from respiratory systems is around 25–30 per million population per day. The consensus of the combined analysis of all available epidemiological studies of PM_{10} is that every $10 \, \mu g \, m^{-3}$ increase in concentration is associated with a 1% increase in deaths. On this basis the 10% increase seen during the 1991 London episode would equate to an exposure of $100 \, \mu g \, m^{-3}$ of PM_{10}, which is consistent with the analysis of the episode. The effects of the episode could have been entirely due to exposure to fine airborne particles.[31,32]

The role of power stations on health in such episodes is now thought to be

[29] Committee on the Medical Effect of Air Pollutants, *Non-biological Particles and Health*, HMSO, London, 1995.

[30] Advisory Group on the Medical Aspects of Air Pollution Episodes, *Sulfur Dioxide, Acid Aerosols and Particulates*, HMSO, London, 1992.

[31] Advisory Group on the Medical Aspects of Air Pollution Episodes, *Health Effects of Exposures to Mixtures of Air Pollutants*, HMSO, London, 1995.

[32] Committee on the Medical Effect of Air Pollutants, *Asthma and Outdoor Air Pollution*, HMSO, London, 1995.

through their contribution to secondary airborne particles.[33,34] These are formed in the atmosphere from gaseous sulfur and nitrogen oxides. Major current air pollution effects have now come full circle and involve processes involved in the formation of acid rain as well as the meteorology associated with episodes of high urban air pollution levels! It should not be concluded that improvements in emission over the past 50 years have not been effective. Instead, it has been recognized that health effects may occur at much lower concentrations than realized formerly. Processes which occurred previously but were obscured by other effects now assume greater importance. With a trend towards smaller gas-fired or alternatively fuelled power stations, the major issues for power station emissions are increasingly nitrogen oxides, the formation of secondary particles of sulfate and nitrate, and the combined contribution of power stations with other types of sources.

10 Local Air Quality Management

To tackle these issues, one needs to know when and where high exposures occur, which requires monitoring, why they occur, which requires information on emissions, and the relation between emissions and concentrations using air quality modelling. These three factors, monitoring, emissions, and modelling, are the essential ingredients of air quality management.

Even with good emission controls, high levels of local traffic or a high density of industrial plants or power stations may lead to local breaches of air quality standards. This is recognized in the 1995 Environment Act, which foresaw a local framework for reviewing and assessing air quality by local authorities, leading to local action plans in addition to national measures. In the UK Government's National Air Quality Strategy for air quality management,[35] it was concluded that nine air pollutants should be included in its local air quality management programme. These are ozone, benzene, 1,3-butadiene, SO_2, CO, NO_2, particles, polycyclic aromatic hydrocarbons (PAHs), and lead. These pollutants roughly correspond with the pollutants under consideration in the European Union's Framework Directive on Air Quality Management.

Local Authority Role

The Environment Act passed in the UK in 1995 requires local authorities to undertake reviews of the air quality in their areas. Following a review, local authorities may designate an area an 'air quality management' area. The authority may then be required to bring forward actions plans to improve the air quality in its area to meet air quality standards, using powers within its control. These powers could range from measures to ensure road vehicles comply with emissions standards, to closing roads to traffic and land use planning measures and reductions in industrial emissions. An example of an air quality review

[33] J. Stedman, *Atmos. Environ.*, 1997, **31**, 2381.
[34] J. Stedman, *Clean Air*, 1998, **28**, 87.
[35] Department of the Environment, *UK National Air Quality Strategy*, Cmnd 3587, Stationery Office, London, 1997.

addressing specifically the combined impact of a number of industrial sources is the assessment of NO_x in the East Thames Corridor.[36]

The air quality review should consider the future air quality by 2005. Defining the accuracy of air pollution modelling results is difficult even in very well tried and tested situations, such as the use of dispersion models for single well-defined sources. One can imagine that dispersion models will be used to define the extent of air quality management areas. For power stations the exceedences in some cases are likely to arise from SO_2 and NO_2. The areas where exceedences are predicted to occur could be fairly widespread. Modern power stations relying on combined cycle gas turbine technology fired by gas (see Table 3 for typical emissions) are unlikely to lead to exceedences of air quality standards on their own. However, the combined effects of road transport in urban areas together with power stations may lead to high predictions locally and the unanswered question of which kind of source should be subject to the most stringent emission controls.

Improvement by National Measures

A review of air quality would take into account improvements in air quality expected in the future. National inventories of PM_{10} emissions are somewhat misleading since the 19% of the national total from diesel vehicles and the 5% of the national total from petrol vehicles will lead to much greater exposures in urban areas compared with the 15% from fossil-fuelled power stations (this is based on the emissions inventory given in EPAQS[37]). Thus in Greater London, the centre of which will undoubtedly become an air quality management area for PM_{10}, 86% by weight of primary PM_{10} emissions are derived from vehicle exhausts.

Reductions in national PM_{10} emissions in the future will arise from the significantly lower emissions from heavy duty vehicles. Reductions will also arise from the increasing number of cars fitted with emission control equipment, which run on unleaded petrol, eliminating the release of lead-rich particles, and reductions in the sulfur content of diesel will reduce particle emissions from diesel vehicles. Although predictions of future national emissions are inevitably uncertain, because of the uncertainty regarding the growth of traffic and the composition of the vehicle fleet,[25,38] they all indicate significant reductions by 2005 of national PM_{10} emissions from road transport of up to 50%. However, reductions could be offset by an increased proportion of diesel cars on the road.

Some of the PM_{10} particles are sulfate and nitrate aerosols, produced in the rural atmosphere from distant sources elsewhere in the UK or abroad, and are not related to urban emissions in London. The sources of this rural background PM_{10} concentration are hard to quantify, but levels will be highest during summer ozone episodes. National measures regarding road transport would not be the main influence. Instead, the influence of national measures to reduce acid

[36] HMIP, *An Assessment of the Effect of Industrial Releases of Nitrogen Oxides in the East Thames Corridor*, HMSO, London, 1993.

[37] Expert Panel on Air Quality Standards (EPAQS), *Particles*, HMSO, London, 1995.

[38] Quality of Urban Air Review Group (QUARG), *Diesel Vehicle Emissions and Air Quality*, Department of the Environment, London, 1993.

Table 3 Typical emissions from different kinds of power stations

Plant	Fuel	Efficiency/%	Emission rate/kg MWh^{-1}			
			CO_2	SO_2	NO_x	Dust
Conventional	Coal	38	900	12.5	4.5	0.5
Conventional with low NO_x and FGD	Coal	37.3	910	1.2	2.5	0.3
Conventional	Oil	37	750	14.2	2.7	0.5
CCGT	Gas	50	400	Negligible	0.5	Negligible

deposition under the various Conventions would be of most significance. This is an interesting example, showing the inter-relationships between the role of the various pollutants. All the 'end of pipe' or 'end of stack' control measures envisaged for road transport or power stations involve some slight loss of efficiency and hence a greater contribution to carbon dioxide emissions (see Table 3). The switch to gas-fired combined cycle gas turbine power stations shows advantages for acid gases and particles on a local and regional scale and for carbon dioxide emissions on a global scale, given adequate supplies of natural gas.

11 Climate Change

Measures to reduce emissions of greenhouse gases have to be considered within a global context. National emission totals are generally a small fraction of global emissions. A protocol on emission reductions in industrialized countries was agreed in Kyoto in 1997, with individual countries within the European Union each agreeing to contribute towards an overall percentage reduction in European Union 1990 greenhouse gas emissions by 2010. Public power (30%) is the single largest contributor to UK greenhouse gas emissions. Other combustion processes (domestic, industrial, commercial) contribute 49%, but it is road transport with a 21% contribution which will come under greatest scrutiny because of its continuing growth. Each sector will be required to make a contribution, but public power in the UK has already contributed the most. Carbon dioxide emissions have fallen from 57 Mt C in 1970 to 44.3 Mt C in 1994. This has arisen as a consequence of the switch to gaseous fuels and the greater efficiency of combined cycle gas turbine plant compared with other types of generation. The striking improvement in CO_2 emissions from the major power companies in the UK is seen in Table 4.

Having stressed in earlier sections the increasingly strong inter-relationship between airborne emissions from power stations and airborne emissions from other sectors within the UK, for greenhouse gas emissions one is even more dependent on the contribution from other countries. Some countries are committed to percentage reductions; developing countries' CO_2 emissions will increase in future years. More than ever before, the power station contribution cannot be judged in isolation and has to be considered as part of an overall economic and environment system. Moreover, the percentage emission reductions

41

Table 4 Average CO_2 emissions from major UK power companies in recent years

Year	1991	1992	1993	1994	1995
CO_2 emissions/kg MWh^{-1}	940	905	850	825	800

are a convenient policy tool but are too small to be associated directly with any measurable climate effect, such as a lower or delayed global temperature rise.

12 Conclusions

It is clear that air pollution remains as much on the public agenda as it did over 50 years ago and power stations are part of this though not necessarily the dominant part. Many similarities exist between action in the 1950s and the 1990s, such as the continued concern over particles and the involvement of local authorities. In absolute terms, progress has been made and air pollution has been reduced. Understanding of the adverse effects of air pollution has improved so that air quality standards have been tightened. Changes have taken place but not always for the reasons foreseen. Remedies have tended to be economic not technical.

At the present time, partly as a consequence of the availability of better monitoring equipment, attention has been paid to a wider range of pollutants. However, the meteorological circumstances under which high concentrations arise are similar to those under consideration 40 years ago. There has been more concern over the combined influence of many sources over distances that were not considered in the 1950s and the inter-relationship between pollutants on a local and regional scale is recognized, even if it is difficult to manage. Regulation of stationary and mobile sources has reduced emissions considerably.

The methodical approach introduced by an air quality management strategy will lead to better understanding of the causes and consequences of air pollution at a local level, but the task involved is not simple. Blanket decision making, regardless of local circumstances, would be easier to implement. It is hoped that the opportunity given to local authorities by the legislation will not be missed.

The availability of information for the public has increased. The designation of air quality management areas will bring strategies for improvement to wider public attention. Plans for improving air quality in these areas will need to be brought forward. From the road transport example considered it may be possible to address general issues satisfactorily, but intractable local problems may remain.

The management of air quality has not been irrevocably fixed by recent policy developments and future changes may be expected. An even more integrated view of pollution is expected to develop in the future. With a trend towards smaller gas-fired or alternatively fuelled power stations, the major issues for power station emissions are increasingly nitrogen oxides, the formation of secondary particles of sulfate and nitrate, and the combined contributions of power stations and other types of sources.

Environmental Performance of the Liberalized UK Power Industry

STEVE ADRAIN AND IAN HOUSLEY

1 Introduction

The privatization of the electrical supply industry (ESI) has led to the unprecedented changes in the UK electricity market. Despite the lack of premium for environmental investments, fierce competition and the adoption of forward-looking environmental policies have resulted in major environmental benefits. Under the pressure of market forces, a major shift from coal-powered generation to gas-powered generation has occurred with the investment[1] in combined cycle gas turbine plant (CCGT). During the same period, 6 GW of flue gas desulfurization plant has been fitted to the most efficient coal-fired plant, cogeneration electrical capacity has increased from 2.3 GW (1991) to about 3.5 GW (1996) out of the Government target of 5 GW by 2000, and 341 MW of wind generation has been built. These major environmental improvements have been undertaken whilst reducing the cost to consumers by about 20% in real terms. This paper deals with the evolution of the ESI in the UK towards sustainability within the new liberalized market. The policies and implementation strategies required to meet the environmental and market challenges are discussed and the response of the industry towards sustainability addressed.

2 Policy

The driving forces for environmental improvement in the performance of the ESI stem from the unique position of the industry. It produces a product, electricity, that is essential to the quality of life of everyone. In delivering this unique commodity to the household or workplace, this product must be generated instantaneously and transported over many miles. At all stages of the process, environmental issues must be addressed. These range from fuel delivery and storage through generation by power stations using fossil, renewable, or nuclear

---

[1] *Electricity Industry Review*, Electricity Association, London, January 1998.

Issues in Environmental Science and Technology, No. 11
Environmental Impact of Power Generation
© The Royal Society of Chemistry, 1999

Figure 1 The evolution of environmental management in the ESI

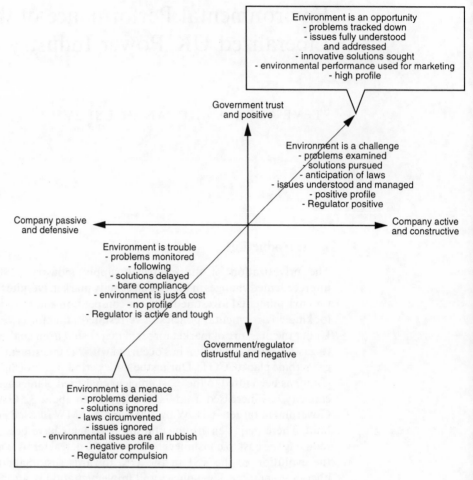

fuels to high voltage transmission and disposal of by-products to market or landfill. In producing and delivering this unique product essential to the quality of life, we must show responsibility to our customers, neighbours, and stakeholders for striking an optimum economic balance between the cost of electricity and environmental impact. A range of policy tools is available to regulators to assist in striking the right balance, *e.g.* the concepts of Best Available Technique Not Entailing Excessive Cost (BATNEEC)[2] and Best Practicable Environmental Option (BPEO),[2] the polluter pays principle,[3] the precautionary principle,[3] and the concept of sustainability.[3] These tools are at the heart of major legislation such as the EU Large Combustion Plant Directive (LCPD),[4] the Environmental Protection Acts (EPA 1990, 1995), the EU Integrated Pollution Prevention and

[2] *Guidance for Operators and Inspectors of IPC Processes, Best Practical Environmental Option Assessments for Integrated Pollution Control*, Environmental Agency, HMSO, London

[3] *Treaty of Rome*, as amended 1973, 1986, 1993, and 1997, Article 130R, HMSO, London.

[4] *EU Large Combustion Plant Directive*, 88/609/EEC and 84/360/EEC, HMSO, London.

Control directive (IPPC),[5] and the EU Air Quality Framework Directive.[6] Furthermore, increasing complexity of regulation and new EU environmental legislation on acidification, water, air quality, heavy metals, ozone, and liability, currently in the draft stage, are likely to have a future impact on the ESI.

To meet the challenges of legislation, to manage relationships with banks and financial institutions, to manage environmental risks, and to demonstrate to our customers that we care for the environment, the ESI has had to move from a defensive position during the 1980s to a position of active management of these important issues. It is instructive to examine the evolution of the ESI (Figure 1) in the UK from the defensive position adopted during the 1980s to the present day. During the period prior to privatization, the industry's stance was to comply with the law and to defend itself with scientific research. Post privatization in 1991 the industry has gradually moved from 'environment is trouble' to a position where it is seen as a challenge and the regulator is regarded as a partner.

To manage environmental risks and to improve reputation, many of the companies within the ESI have adopted new environmental policies and have developed and implemented environmental management systems (EMS). National Power's policy statement[7] below provides an example of such policies:

Environmental Policy Statement

- To integrate environmental factors into our business decisions wherever we operate:
  - By making cost effective investments which continue to improve our environmental performance
  - By assessing, managing and controlling environmental risks associated with our current and planned activities
- To monitor compliance with environmental regulations and, where appropriate, to perform better than they require:
  - By continuing to control and reduce emissions
  - By establishing clear measurable environmental targets across the Company, including sustainability
- To improve our environmental performance continuously:
  - By raising employees' awareness so that they can carry out their environmental responsibilities effectively
  - By demonstrating our commitment to sound environmental management
  - By minimizing environmental incidents and complaints
  - By using energy, materials, and natural resources efficiently and in a more sustainable way
  - By promoting the adoption of good environmental management practices by our contractors and suppliers
- To review regularly at Board level, and to publish our environmental performance:

---

[5] *EU Integrated Pollution Prevention and Control Directive*, 96/61/EC, HMSO, London.

[6] *EU Air Quality Framework Directive*, 96/62/EC, HMSO, London.

[7] National Power, *Environmental Performance Review 1998*, National Power, http://www.national-power.com.

- By annually reviewing the implementation of policy
- By setting high standards in open and transparent reporting at a Corporate and Operational level
- To maintain our reputation for effective environmental management:
  - By remaining in the forefront in the employment of best practice environmental management systems
  - By establishing and maintaining effective environmental management systems consistent with ISO 14001 and, where appropriate, EMAS (EU Eco-Management and Audit System)
  - By developing a positive and constructive relationship with local communities, regulators, and authorities

Examination of the policies of other companies in the ESI shows some variations in emphasis on the policy outlined above to include:

- Promotion of the development of more efficient and cleaner ways of carrying out operations and activities
- To manage land resources with sensitivity and to promote and conserve natural habitats and heritage
- To promote research and development into environmental effects

These policies address the requirements on the industry to continuously improve its environmental performances by:

- Balancing environmental protection and economic benefits in optimizing major capital investments in a way so as to avoid expensive retrofits
- Economically and environmentally optimizing investments through the life of a major asset
- Optimizing the revenue expenditure against environmental benefits from the process
- Regularly assessing the environmental risks and updating the management plans
- Managing long term liabilities

3 Policy Implementation—ISO 14001 and EMAS

For an effective environmental management system (EMS) it is essential for the environmental policy to include commitments to prevention of pollution, commitments to compliance, and to pay due regard to preventative action. Furthermore, commitment is required from the top of the company through reviews at board level of policy. In addition, the policy should clearly state the roles and accountabilities of staff and executives in implementation of the system. A typical list of roles and accountabilities is shown in Table 1.

To implement policy it is necessary to implement an effective EMS to ensure that all environmental risks are understood, monitored, managed, controlled, and audited in order to set standards for improved performance against the company's environmental objectives and targets.

Table 1 Roles and Accountabilities within the EMS

	Policy accountabilities	Business role	Implementation role	Review role
CEO	Ultimate 'duty of care'	To understand the impacts on commercial position and stakeholder image		
Executive	Sets policy and reviews annually	To understand the impacts on commercial position and stakeholder image		
Environmental Policy Manager	Develops policy and strategy for implementation	To analyse the impacts on the business and to interpret and develop responses for dissemination throughout the company	To monitor and audit implementation across the company, using functional support as appropriate	To review policy, implementation, operation and compliance across the businesses
Corporate Managers		To understand the impacts on their functional responsibilities and business processes	To implement as appropriate within their corporate responsibilities	To review implementation and operation of policy as it impacts on their corporate responsibilities
Station/site Managers	Site duty of care	To understand the impact on operational responsibilities and business processes	To implement as appropriate to their operational and site responsibilities	To review implementation and operation of policy as it effects operational and site activities
Staff		To understand impact within their individual responsibilities	To implement as appropriate within their individual responsibilities	

Figure 2 Environmental
management system cycle

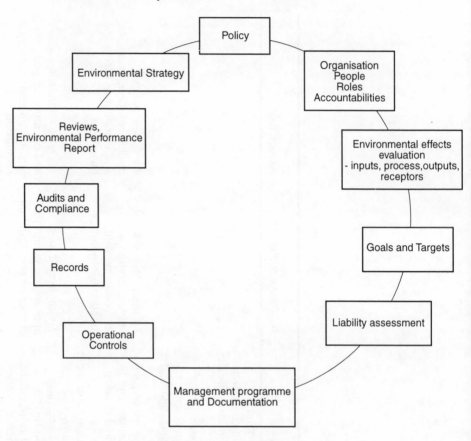

There is a variety of systems in use by major companies to manage environmental risks, ranging from those based on the quality standard ISO 9001 to those based on ISO 14001 and EMAS. In our opinion the ISO 14001 scheme is more appropriate than the quality standard ISO 9001 since it more readily lends itself to a change of culture from compliance to continuous improvement beyond compliance. ISO 14001 does not need to be a bureaucratic system and can be implemented in a way that supports a company's continuous improvement culture without imposing a high management burden at operating sites. An EMS based on ISO 14001 is illustrated in Figure 2.

Once policy and the organization, roles, and accountabilities are in place, a thorough examination of all elements of environmental (environmental effects evaluation) risk is undertaken. Depending on the scale of the risk and the economic consequences, the risks are prioritized and goals and targets selected for improvement of performance. The goals and targets are then incorporated into a management plan and appropriate procedures are drawn up for managing and reducing the risk. In National Power we assess potential liabilities at this stage and these are incorporated in the management plan and in the performance standards for those managers managing the risk. The management plan will cover targets and goals for improvements through investments and enhanced operational control. The plant will be managed and operated in accordance with the plan. To ensure compliance and continuous improvement, regular self-audits are initiated as well as major audits covering engineering risk, finance, health and safety, and environmental compliance. To ensure that performance is monitored, all environmental data are collected, validated, and verified. These verified data are used in the company's Environmental Performance Report (EPR) in which progress is reported against sustainability objectives and targets. In the case of National Power, local reports compatible with the principles of the EU Eco-Management and Audit Systems (EMAS)[8] are produced. At this stage in the process the company's environmental strategy and policy is reviewed. Views of interested parties are sought in reviewing the process to ensure ownership and effective communication. In this way, all staff are aware of their targets, objectives, and performance and the executive are involved in reviews of audits, EPR, environmental strategy, and policy. Of course, all staff need to be trained and re-trained in the operation of the system and in understanding the environmental risks.

The EMAS scheme came into effect in the UK in April 1995 and National Power's Drax Power Station was the first power station to achieve certification in the UK. In practice, there is little difference between ISO 14001 and EMAS. In addition to the requirements of ISO 14001, the latter has the requirement to carry out systematic audits at intervals of not more than three years, to have the statements validated by independent accredited verifiers at the time of audit, and to submit the statements to the appointed competent body and make them available to the public.

In order to manage and control operational risk more rigorously, a number of companies, including Baxter (USA), Monsanto (USA), and Statoil (Norway), are working on environmental accounting systems to account for all revenue and capital expenditure incurred in managing the environmental risks. National Power and others such as PowerGen and Eastern Group already account for all emissions and capital expenditure. Within the ESI, trials are taking place at Yorkshire Electricity and National Power to develop comprehensive environmental accounting systems.

4 Meeting Legislative Requirements

Recognizing the UK Government's commitments under the LCPD Directive, Kyoto, the Second Sulfur Protocol, and agreements with regulators together with the concerns over acidification, particulates, and greenhouse gases, the industry has invested in a series of improvements to meet current and likely environmental limits. These investments, over £2 billion at National Power alone, have been in abatement technologies at coal-fired power stations, replacement of old coal plant by modern, high-efficiency gas-fired plant, investment in gas-fired cogeneration plant, and in renewables such as wind farms. In carrying out these environmentally driven investments, the ESI has succeeded in reducing the cost to the consumer by 18–20% in real terms (1990–1997) whilst improving the diversity of fuel supply and modernizing the plant portfolio.

---

[8] *EU Eco-Management and Audit System*, EC 1836/93, DETR, London.

Table 2 Abatement costs for high utilization plant

Abatement technique	t^{-1} removed	Capital cost/$ kW^{-1}$
FGD	400–550	90–300
Combustion control of NO_x	20–50	50–2000
SCR	1600–3500	70–140
Particulate control	? very plant specific	50–110

Abatement Technologies

The main abatement technologies that have been installed are flue-gas desulfurization (FGD) at National Power's 4 GW Drax Power Station and PowerGen's 2 GW Ratcliffe Power Station. Recognizing that FGD is BATNEEC and BPEO for high utilization coal plant, both companies installed the limestone–gypsum process on the most efficient of their coal-fired plant. Drax Power Station has achieved over 90% reduction of SO_2 with 99% FGD plant availability. The costs of abatement are very dependent on fuel specification, by-product disposal costs, space constraints, plant design, site constraints, plant unit size, plant supply market, and the size of the power island. Table 2 shows the marginal costs of abatement for high utilization plant. The costs rapidly rise for low utilization, short-life plant, and given that the liberalized market in the UK allows no premium for abated plant, FGD is clearly not BATNEEC for such plant. Similarly, Table 2 shows the marginal cost of abatement for NO_x. It can be seen that the selective catalytic reduction techniques (SCR) is not considered BATNEEC for retrofitting to current plant. Instead, the abatement technology of choice has been the installation of low NO_x burners to achieve about 50% reduction in NO_x emissions from coal- and oil-fired power plant.

Gas-fired Plant Investment

Given the environmental risks, government international commitments, and the availability of high-efficiency gas plant, the ESI has invested in about 17 GW of new CCGT out of a total capacity of about 64 GW. State-of-the-art CCGT plants, such as Didcot B with an efficiency of approximately 56%, offer major environmental improvements such as almost no SO_2, low NO_x, no particulates, and low water usage.

5 Impact of New EU Legislation and International Standards

Revision of the Large Combustion Plants Directive

The new Large Combustion Plants Directive as currently drafted will have major impacts on new power projects in the European Union from the year 2000 onwards. In the case of CCGT plant, the Directive will ensure that the latest low NO_x burner technologies are used. In designing the emission limit, allowance is made for efficiency, thus encouraging investment in high performance plant. It is, however, disappointing to see that the principles used in setting limits for coal-, oil-, and gas-fired boilers continue to be based on best available technology grounds rather than BATNEEC. Moreover, there are no economic or environmental reasons why different standards are proposed for coal, oil, gas, and biomass fuels. There is also no reason why the limits should be independent of oxygen reference conditions. This, in effect, unfairly penalizes plant fired on liquid and gaseous fuels (3% O_2) as compared with solid fuels (6% O_2). It is also difficult for plant operators to understand why the proposed NO_x limit for gas-fired conventional boilers is more onerous than the limit for CCGT. No justification is given. Another demerit of the Directive is that no account is taken of utilization. Furthermore, proposals are made for dealing with issues that are better dealt with by the appropriate competent authority of each of the member states. For example, the monitoring and breakdown provisions are inappropriate in a deregulated market since they fail to take into account the energy policy, fuel supply, and economic circumstances that prevail in the member states. In the case of the breakdown provisions, only 120 hours is allowed to rectify the failure of abatement plant. Many credible failure modes, such as the plant-type fault at Drax during 1998, require much longer to rectify and if the conditions in the Directive are applied, energy from such high efficiency plant would have to be replaced by energy from lower efficiency and often 'dirtier' plant.

The revised emission values for new plants (authorizations granted after 1/1/2000) mean that any new 'black fuelled' plant will require SCR for NO_x control to meet the standard of 200 mg m^{-3} and FGD for SO_2 control to meet the standard of 200 mg m^{-3}.

Air Quality

New legally binding air quality standards[6] on SO_2, NO_x, particulates, and lead are to be introduced from the year 2005 onwards (Tables 3–7 summarize the standards). Consequently, it is likely that the standard for SO_2 will have an impact on large 'black-fuel' plants that are not fully abated.[9] It may well be necessary for the plant operator to change the fuel specification or to make further investments. In the case of NO_x it is unlikely that power plant would be required to take action to meet the standards.[9]

UNECE Heavy Metals Protocol

The provisions of the draft UNECE Heavy Metals Protocol appear to be in line with ESI expectations, so current or planned investments should allow these standards to be fully met.

World Bank Standards

World Bank Environmental standards are currently under revision. The last draft was issued in September 1997. These standards form a benchmark for all

---

[9] D. Laxen, *Generating Emissions, Studies of the Local Impact of Power Generation*, National Power, Swindon, 1996.

Table 3 Ambient air concentrations (limit and guideline values for effects of SO_2 on health)

Limit or guideline	SO_2/ppb	Averaging period and/or compliance level
EC Directive 80/779 Limit Values		
If median daily smoke $< 40\,\mu g\,m^{-3}$	45[a]	Median of daily means over a year
If median daily smoke $> 40\,\mu g\,m^{-3}$	30[a]	Median of daily means over a year
If 98%ile daily smoke $< 150\,\mu g\,m^{-3}$	132[a]	98%ile of daily means over a year and on no
If 98%ile daily smoke $> 150\,\mu g\,m^{-3}$	94[a]	more than 3 consecutive days
If median daily smoke $< 60\,\mu g\,m^{-3}$	67.5[a]	Median of daily means over the winter period
If median daily smoke $> 60\,\mu g\,m^{-3}$	49[a]	Median of daily means over the winter period
EC Directive 80/779 Guideline Values	15–22.5[a]	Annual mean
	38–56[a]	Daily mean
EU Directive 96/62 (Proposals) (by 2005)	131.1[b]	99.7%ile of hourly means over a year
	46.8[b]	99.2%ile of daily means over a year
WHO Guideline Values (1997)	175[c]	10-minute mean
	43.7[c]	Daily mean
	17.5[c]	Annual mean
WHO Guideline Value (1987)	122[c]	Hourly mean
UK National Air Quality Strategy (Proposal) (by 2005)	100	15-minute mean (EPAQS standard) in $\geq 99.9\%$ of 15-minutes in a year
DETR Air Quality Bandings		
'Low'	< 100	15-minute mean
'Moderate'	$\geq 100 - < 200$	15-minute mean
'High'	$\geq 200 - < 400$	15-minute mean
'Very high'	≥ 400	15-minute mean

[a]Converted from $\mu g\,m^{-3}$ values assuming 293 K and 101.3 kPa, *i.e.* a factor of 2.66.
[b]Converted from WHO $\mu g\,m^{-3}$ values by EU assuming 293 K and 101.3 kPa, *i.e.* a factor of 2.66.
[c]Converted from WHO $\mu g\,m^{-3}$ values by WHO assuming 273 K and 101.3 kPa, *i.e.* a factor of 2.86.

international projects, and therefore changes to them must affect the approach of international power producers to the development of power projects.

EU Waste Incineration Directive

The latest draft of the EU Waste Directive does not differentiate the risks from different waste streams, making Waste Combustion Schemes, and co-firing waste with coal or oil, potentially very expensive. There is also a risk that fuels such as

Table 4 Limit and guideline values for effects of NO_2 on health

Limit or guideline	NO_2/ppb	Averaging period and/or compliance level
EC Directive 85/203 Limit Value	104.6[a]	98%ile of hourly means over a year
EC Directive 85/203 Guideline Values	70.6[a]	98%ile of hourly means over a year
	26.2[a]	Median of hourly means over a year
EU Directive 96/62 (Proposals) (by 2010)	104.6[b]	99.8%ile of hourly means over a year
	20.9[b]	Annual mean
WHO Guideline Values (1997)	104.6[c]	Hourly mean
	21–26[c]	Annual mean
UK National Air Quality Strategy (Proposals) (by 2005)	150	Hourly mean (EPAQS standard)
	21	Annual mean
DETR Air Quality Bandings		
'Low'	< 150	Hourly mean
'Moderate'	≥ 150– < 300	Hourly mean
'High'	≥ 300– < 400	Hourly mean
'Very High'	≥ 400	Hourly mean

[a]Converted from $\mu g\,m^{-3}$ values assuming 293 K and 101.3 kPa, *i.e.* factor of 1.91.
[b]Converted from WHO $\mu g\,m^{-3}$ values by EU assuming 293 K and 101.3 kPa, *i.e.* a factor of 1.91.
[c]Converted from WHO $\mu g\,m^{-3}$ values by WHO using a conversion factor of 1.91, *i.e.* assuming 293 K and 101.3 kPa.

petcoke could be classified as a waste. If implemented, co-firing waste with coal or oil is likely to become too expensive.

Sulfur in Liquid Fuels Directive

Ministers agreed at the Environmental Council in June 1998 to the 'Sulphur in Liquid Fuels Directive', which limits the sulfur in heavy fuel oil to 1% by 2003. This is likely to increase the price of such oil and hence the cost of electricity from oil-fired power stations.

EU Acidification Strategy

In accordance with the EU Fifth Environmental Plan, an 'Acidification Strategy' has been drawn up using an optimization model. The intention of the Strategy is to greatly reduce the area of the eco-system over which critical loads are exceeded. To date the model has used cost data which are in error and fail to recognize alternative investments or the impact of plant utilization on cost.

Table 5 Limit and guideline values for effects of O_3 on health

Limit or guideline	O_3/ppb	Averaging period and/or compliance level
EC Guideline Value	55[a]	8-hour mean
EC threshold Values		
Public information	90[a]	1-hour mean
Public warning	180[a]	1-hour mean
WHO Guideline Value (1997)	60[b]	Running 8-hourly mean
UK National Air Quality Strategy (Proposal) (by 2005)	50	Running 8-hour mean (EPAQS standard) on $\geq 97\%$ of days in a year
DETR Air Quality Bandings		
'Low'	< 50	Running 8-hour mean
'Moderate'	$\geq 50-<90$	Hourly mean
'High'	$\geq 90-<180$	Hourly mean
'Very High'	≥ 180	Hourly mean

[a]Converted from $\mu g\,m^{-3}$ values assuming 293 K and 101.3 kPa, *i.e.* factor of 1.99.
[b]Converted from WHO $\mu g\,m^{-3}$ values by WHO using a conversion factor of 1.99, *i.e.* assuming 293 K and 101.3 kPa.

Table 6 Limit and guideline values for effects of PM_{10} on health

Limit or guideline	$PM_{10}/\mu g\,m^{-3}$	Averaging period and/or compliance level
EU Directive 96/62 (Proposals)	50	90.4%ile of 24-hour means over a year by 2005 (98% by 2010)
	40	Annual mean by 2005 ($20\,\mu g\,m^{-3}$ by 2010)
WHO Guideline Value (1997)	none	
UK National Air Quality Strategy (Proposal) (by 2005)	50	Running 24-hour mean (EPAQS standard) on $\geq 99\%$ of days in a year
DETR Air Quality Bandings		
'Low'	< 50	Running 24-hour mean
'Moderate'	$\geq 50-<75$	Running 24-hour mean
'High'	$\geq 75-<100$	Running 24-hour mean
'Very High'	≥ 100	Running 24-hour mean

Furthermore, the model does not adequately deal with the major uncertainties in critical loads (up to factor of 4), in modelling of dispersion, and deposition (factor of 2), nor does it justify an implementation timescale of 2010. If current proposals come into force, then huge sums of money will be spent in protecting at most

Table 7 Guideline and recommended values for effects on vegetation

Guideline or recommendation	SO_2/ppb[a]	NO_2/ppb[b]
EC Directive 96/62 Guideline Values (Proposals)		
Annual mean	7.5[c]	15.7[d]
WHO Guidelines		
Crops (annual mean)	10.5[c]	
Forest and natural vegetation		
Annual mean	7.0[c]	15.7[d]
Daily mean	35	
4-hour mean		50.5
Sensitive forest and natural vegetation (annual mean)	5.2[c]	5.3
Lichens (annual mean)	3.5[c]	
Sphagnum-dominated (annual mean)		6.2
IUFRO Guidelines for Normal Woodland		
Annual mean	17.5	
Max. no. days > 35 ppb		
Winter	12 days	
Summer	12 days	
97.5%ile of 30-minute means	52.5	
IUFRO Guidelines for Sub-Optimal Woodland		
Annual mean	8.7	
Max. no. days > 17.5 ppb		
Winter	12 days	
Summer	12 days	
97.5%ile of 30-minute means	26	
UNECE Guidelines		
All (annual mean)		15.7[d]
(4-hour mean)		49.7[d]
Agricultural crops (annual mean)	10.5[c]	
Natural vegetation (annual mean)	7[c]	
Forest (annual mean)	7[c]	
Cyanobacterial lichens (annual mean)	3.5	

[a]Converted from $\mu g\,m^{-3}$ values using conversion factor 2.86, as stated in WHO guide, *i.e.* 273 K and 101.3 kPa.
[b]Converted from $\mu g\,m^{-3}$ values using conversion factor 1.88, as stated in WHO guide, *i.e.* 298 and 101.3 kPa.
[c]Annual and winter means.
[d]$NO_2 + NO$.

1–3% of eco-systems. For example, in the UK the marginal cost of abatement for implementation of the Strategy to SO_2 would be in excess of £5 M kt^{-1} a^{-1}. European industry is pressing the EC to develop a better strategy with better cost data applied to relevant market models and using better scientific data.

Global Warming

EU environment ministers agreed on 17 June 1998 to a new 'burden sharing' arrangement with regard to the EU's commitment under the Kyoto protocol to reduce greenhouse gas emissions by 8% from 1990 levels by 2008–2010. The

UK's commitment has been increased from the EU's original proposal (linked to its Kyoto negotiating stance of 10% total EU reduction) from a 10% reduction (on 3 greenhouse gases) to 12.5% (on 6 gases).

The UK ESI has invested heavily in low-carbon processes and hence most of the base-load generation now comes from nuclear and gas plant. It will be much harder and much more expensive to convert low utilization plant to low-carbon processes. Moreover, further conversion to gas will threaten the diversity of fuel requirement in the recently announced UK Government's Energy Policy document. The Electricity Association in the UK has conducted a project on the potential for significant, cost-effective reductions in carbon at the point of consumption and concludes that up to 35.4 Mt C per annum can be saved for an investment of £20 billion in energy efficiency measures in the domestic, commercial, industrial, and transport sectors. The largest contributing sector would be the domestic sector with a 6 Mt C per annum saving at a cost of £11 billion.

6 Environmental Performance Post Privatization

As a result of the implementation of policy, there have been substantial reductions in the emissions to the environment from the UK electricity generating sector. As stated previously, the drivers for these reductions have been a complex mixture of both commercial/environmental business management and environmental legislation.

The previous section considered the impact of EU environmental legislation on the UK power generation industry. It is not always the case that the UK follows the EU. In many instances the UK has set the standards; for example, Integrated Pollution Control (IPC) Part 1 of the Environmental Protection Act 1990 is just being implemented in the EU with the recent Integrated Pollution Prevention and Control Directive.[5]

The UK power industry is regulated under IPC and hence each operating site has to operate under the constraints specified within the Environment Agency's authorization. It is worth noting that IPC came into force at the same time as the industry was being privatized and therefore the improvements in environmental performance post privatization are the result of what could be described as the synergy between these two events.

It is not possible in today's commercially driven electricity market to succeed without due regard for environmental performance. The main UK generating companies have interests in UK and overseas power generation plants. Financing developments and acquisitions always involves making environmental statements in terms of how projects will be developed, managed, and operated. Moreover, lending institutions look at environmental performance in their assessment of financial risk. The environmental performance of a company in its home country is often used as a key indicator, based on the assumption that 'if you can't manage it at home what chance have you anywhere else?'.

The 'City' always reacts to major environmental incidents by a downward trend in share prices. Of course, it need not be a major incident since it is often perception that governs the response of both the media and the general public. However, perception is usually based on a company's past performance.

Table 8 Emissions to air

Year	Electricity delivered to final users/ TWh	CO_2 Mass emission/ Mt	CO_2 Emission per kWh/ $g\,kWh^{-1}$	SO_2 Mass emission/ kt	SO_2 Emission per kWh/ $g\,kWh^{-1}$	NO_x Mass emission/ kt	NO_x Emission per kWh/ $g\,kWh^{-1}$
1990	264.1	198	54	750	2.722	10.31	2.96
1991	270.4	198	54	732	2.534	9.37	2.51
1992	271.2	187	51	690	2.433	8.97	2.48
1993	276.1	169	46	612	2.096	7.59	2.10
1994	275.2	161	44	585	1.764	6.41	1.92
1995	282.4	161	44	570	1.588	5.62	1.76

Emission Reductions

Power stations emit to all three media, *viz.* air, land, and water. However, although power stations, especially coal-fired stations, emit relatively high masses of certain pollutants, the actual impact on local air quality is relatively low due to the tall stacks used and the dilution in the atmosphere of these emissions. If we consider sulfur dioxide emissions, then the contribution above background of a plant is usually only a few ppb.

Table 8 shows the emissions of the three principal gases released into the atmosphere by power stations for the years 1990–1995, *i.e.* CO_2, SO_2, and NO_x, the reductions in terms of both mass and per unit of electricity generated (kWh), and the electricity delivered to final users.[1] These reductions have been achieved through the improvements in thermal efficiency, *i.e.* less emission per unit of electricity generated, the investment in and improvements to abatement equipment such as the Flue Gas Desulfurization plants (FGD) at National Power's Drax power station and PowerGen's Ratcliffe power station. The move to gas-fired Combined Cycle Gas Turbines with typical efficiencies of 50–60% and negligible sulfur emissions has significantly improved the environmental performance of the fossil-fuelled generation sector of the ESI.

Since 1990 the UK has reduced its emissions of carbon into the atmosphere by 11 Mt. Of this reduction, the power generation industry has contributed a reduction of 14 Mt which has been partly offset by increases in emissions from other sectors, in particular from the transport sector.

The UK Government's National Plan for reductions of SO_2 and NO_x for implementing the EC Large Combustion Plant Directive[4] has set the power industry increasingly stringent targets for these gases. However, the ESI has responded to the challenge so well that by 1995 the emissions from power stations were:[1]

- SO_2 reductions—32% better than target
- NO_x reductions—34% better than target

Reductions in emissions and discharges have not only been achieved for air.

57

Coal-fired power stations produce large quantities of ash, consisting of pulverized fuel ash (PFA) and furnace bottom ash (FBA), which is disposed of either to landfill sites or to market as a useful by-product for the aggregates and construction industry. Every tonne of ash sold to market saves the equivalent amount of primary aggregate extraction. In the 1995/96 financial year, 1.8 Mt of PFA and 2.1 Mt of FBA were sold; however, owing to conditions in the construction industry, 4.4 Mt of PFA and 0.1 Mt of FBA had to be sent to landfill.[10] As soon as it becomes economic to sell to markets, the industry will recover PFA from landfill for sale.

7 Towards Sustainability

The accepted definition for sustainability is the Brundtland definition[11] as follows:

- Development that meets the needs of the present without compromising the ability of future generations to meet their own needs

Furthermore, the UK Roundtable on Sustainable Development has (December 1997) identified key indicators of sustainable development which cover the five areas of:

- Consumption of non-renewable resources
- Pollution of air, water, and land
- Social issues
- Biodiversity
- Landscape and cultural amenities

Applying this definition to the ESI means that the industry must take care, wherever economic, to produce electricity with optimum use of primary energy resources, to reduce emissions, to continue to invest in cleaner plant and renewables, to continue to keep the cost of electricity low to ensure that all our citizens improve their quality of life, and to ensure that we manage our operations in such a way as to improve biodiversity and landscape and cultural resources.

Since privatization, the ESI environmental policies have led to a transformation in the environmental performance of the industry. For example, Figure 3 shows the changes in primary fuel use in the industry. This transformation has converted the baseload to low carbon, high efficiency processes which has allowed the UK Government to meet its Rio targets. The industry now uses less primary energy to manufacture electricity, thereby saving that resource for future generations. In the case of gas, the development of new technologies exploiting the advances in the aero-industry has led to efficiency improvements from about 25% to about 60% for CCGT plant, again saving primary fuel for future generations.

The industry has also invested in renewable energy sources such as wind power, hydro-electric schemes, and developed Combined Heat and Power (CHP) schemes. CHP schemes utilize both the heat and the power produced from the

---

[10] *Digest of UK Statistics*, HMSO, London, 1997.
[11] G. H. Brundtland, *Our Common Future*, World Commission on Environment and Development, 1987.

Figure 3 Changes in fuels used for power generation

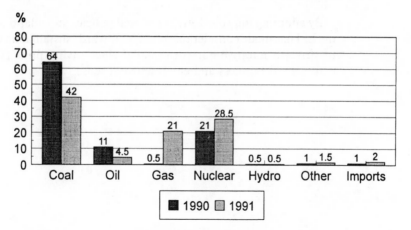

combustion of fossil fuels with overall cycle efficiencies in the order of 80%. This equates to a 35% reduction in primary energy usage and more than 30% reduction in CO_2 emissions.[1] Since privatization there has been a substantial increase in the number of schemes, approximately 100 per year, and by 1996 there were 1336 sites with CHP, amounting to an installed capacity of 3562 MW (electrical), which equates to 6% of the total electricity used by final users and 17% of electricity consumed by industry.[1]

The use of renewable sources, although a relatively small proportion of the UK's power supply (in 1996 this was 1.7%), is developing, assisted by the Non Fossil Fuel Levy (NFFL) promoted by the UK Government. Owing to these initiatives, renewable generation projects with a total capacity of about 2.3 GW have received support. These cover proposals for a range of sizes of scheme up to in excess of 100 MW and covering hydro, landfill gas, municipal waste, sewage gas, wind, CHP, *etc.* We expect that further investment will be necessary if the Government is to meet its target of 10% renewable contribution to electricity needs by the year 2010.

The industry has also been active in promoting energy efficiency and it is expected that the £76 M spent on 380 projects during the last three years will lead to a saving of about 2 Mt C over the lifetime of the schemes.

With regard to the social element of sustainability, the industry has reduced costs, allowing more people to improve their quality of life whilst improving standards of performance in disconnections to customers (the average time lost per connected customer has improved from 35.5 min in 1991/2 to 16.3 min in 1996/7), in system availability (average system availability in 1991/2 was 92.5% in comparison to 95.83% in 1996/7), and in disconnections to domestic customers which have now fallen to 471 (1996/7) compared with over 80 000 in 1991/2.

At National Power we now have amenity maps for all our operational plant to ensure that we manage our sites in ways which take regard of the archaeological, landscape, and amenity aspects of the land. Similarly, we are developing metrics for biodiversity in order to improve the habitats provided by our sites. These initiatives are already bearing fruit, with flourishing colonies of the great crested newt at our Didcot Power Station and peregrine falcons nesting in the chimney stacks of our plants at Fawley and Littlebrook.

By adopting improved environmental policies and adapting to market forces, the ESI has made considerable progress towards sustainability. We will continue to strive to improve our performance since saving primary energy resources makes good business and environmental sense.

BPEO Approaches to the Design and Siting of Power Stations

COLIN POWLESLAND

1 The BPEO Concept

The term Best Practicable Environmental Option (BPEO) was introduced by the Royal Commission on Environmental Pollution (RCEP) in their Fifth Report on Air Pollution Control in 1976,[1] where it was seen as an extension to the concept of Best Practicable Means (BPM). The BPEO concept was further developed by the RCEP in their 10th and 11th Reports[2,3] and amplified in their 12th Report[4] (Best Practicable Environmental Option) in 1988. Although the concept developed from the need for improved coordination of pollution control, it has wider application in that it requires a systematic approach to decision making in which the practicality of all reasonable options is examined and environmental considerations play a major role in determining the final decision. The RCEP[4] define the BPEO as:

'The outcome of a systematic consultative and decision-making procedure which emphasizes the protection and conservation of the environment across land, air, and water. The BPEO procedure establishes, for a given set of objectives, the option that provides the most benefit or least damage to the environment as a whole, at acceptable cost, in the long term as well as in the short term'.

The objective of the BPEO is to achieve reductions in environmental pollution

---

[1] Royal Commission on Environmental Pollution, *Air Pollution Control: An Integrated Approach*, Fifth Report, Cmnd. 6371, HMSO, London, 1976.
[2] Royal Commission on Environmental Pollution, *Tackling Pollution—Experience and Prospects*, Tenth Report, Cmnd 9149, HMSO, London, 1984.
[3] Royal Commission on Environmental Pollution, *Managing Waste: The Duty of Care*, Eleventh Report, Cmnd 9675, HMSO, London, 1985.
[4] Royal Commission on Environmental Pollution, *Best Practicable Environmental Option*, Twelfth Report, Cmnd 310, HMSO, London, 1988.

Issues in Environmental Science and Technology, No. 11
Environmental Impact of Power Generation
© The Royal Society of Chemistry, 1999

and improvements to the quality of the environment as a whole, taking into account the economic implications of different options. The features of a BPEO study on power station design and siting will include the following:

(i) A wide ranging and imaginative consideration of alternative options for power generation, siting, and pollution control options.

(ii) Consideration of both local and long-range effects over both short and long timescales to all environmental media throughout the life cycle of the process (*i.e.* during construction, operation, and decommissioning). The assessment should take into account the present stage of knowledge of available technology in the generation sector concerned together with the scientific understanding of any environmental impacts which may arise.

(iii) Consideration of the financial implications should include both capital and operating costs borne by the developer and, where appropriate, external costs borne by local communities, external organizations or the public purse.

(iv) Determination of the BPEO requires that a reasonable balance is struck between the overall costs and benefits. The RCEP[4] make it clear that where local social factors or political considerations lead to a different decision from that which would have been arrived at by consideration of the environmental impact and costs alone, the outcome should not be termed the BPEO and the basis for the judgement clearly recorded.

(v) A basic principle of the BPEO procedure is the consideration of impacts across all environmental media, whether direct, indirect, or arising from migration between media. However, in many cases, proper quantitative assessment of impacts will not be possible and more qualitative estimates or value judgements applied. As a result, it is important that the procedure for the selection of the BPEO is fully accountable and an audit trail is properly maintained.

2 BPEO Assessment Procedure

A wide range of power generation technologies are available, from tidal barrage schemes, wind farms, gas fired turbines, and coal fired stations to nuclear stations and advanced techniques such as fuel cells. Each will have its own specific set of costs and benefits and some will be more appropriate for particular locations and generating markets than others. It is not possible to provide detailed specific guidance on appropriate assessment techniques for all the technologies currently available; instead, a more general approach to the problem is suggested.

A BPEO assessment might be undertaken at either or both of two stages in the development process:

(i) Initially, to inform the selection of a suitable site and choice of process type. This would be a wide ranging study, both geographically and technically, carried out largely on the basis of engineering calculations. The outcome would be the selection of a particular combination of location and broad process type, together with a list of issues to be resolved at stage (ii).

(ii) Secondly, as part of the detailed design stage once the location and an outline of the process were known, to address the issues identified during the initial assessment and identify in detail the BPEO for the process at that location.

Undertaken in this way, there are many features of the BPEO procedure which overlap the requirements of planning and pollution control legislation and their respective regulatory regimes. Work undertaken at this stage can therefore provide a basis for later submissions to appropriate regulatory authorities. The discussion in this and following sections is related primarily to the first of the stages, *i.e.* selection of location and process type. However, the principles can readily be extended to the selection of the detailed site-specific BPEO.

The BPEO process is outlined in Figure 1 and consists of a number of defined stages. Throughout the process is the requirement to maintain an audit trail. This enables the BPEO procedure to be open and amenable to legitimate review by third parties, thereby providing a robust basis for any future investment decision. Moreover, the audit trail will allow the checking of alternative assumptions and judgements should new information become available. There can be no set procedures for maintaining the audit trail, given the wide variety of projects which will be assessed; however, a number of principles can be identified:

- Primary data sources should be referenced
- Assumptions and judgements should be documented; the basis for the decision should be recorded, together with the names and positions of individuals or groups responsible
- Methods used to generate secondary information from primary information should be recorded and justified
- The basis of any criteria used to compare options should be clearly reported together with the outcome of any assessment. This is particularly important where disparate measures are being compared, for example environmental impact *versus* costs
- In order to facilitate third party examination or even cross checking once the original project team has dispersed, the flow of information, procedures, and decisions should be illustrated by means of a suitable diagram. Information within the audit trail can then be referenced to match the stages in the diagram

3 Assessment Methodology

Definition of Objectives

The assessment needs to commence with a clear statement of the objectives of the study; for example, to identify the location and plant type which represents the BPEO for the generation of electricity by company 'X'.

Ideally, the BPEO approach should be used to identify both location and plant type without any preconceptions as to the outcome. However, this is unlikely to be realized in practice as the developer will usually have identified a particular market opportunity, and flowing from this will be a number of economic,

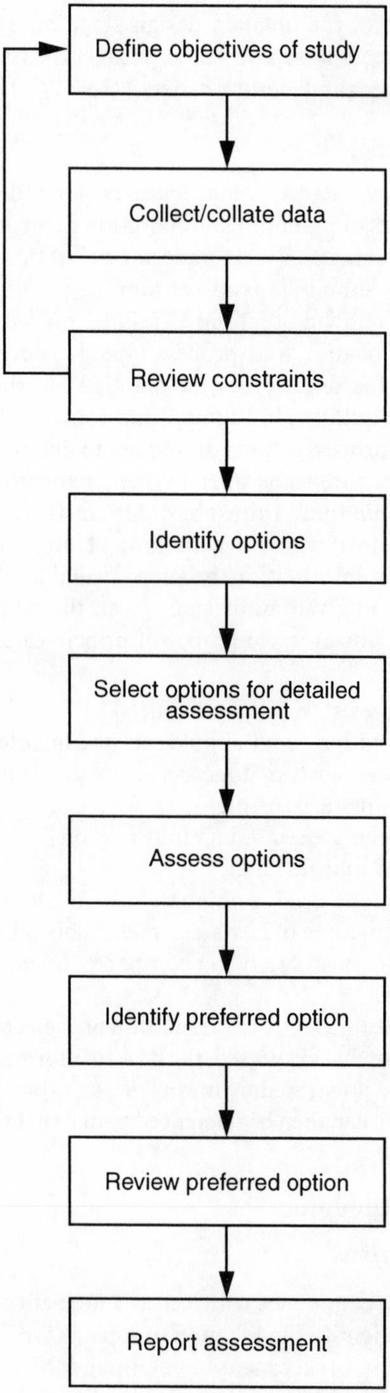

Figure 1 BPEO
assessment methodology

```
        ┌──────────────────────────┐
    ──▶ │  Define objectives of study │
   │     └──────────────────────────┘
   │                  │
   │                  ▼
   │     ┌──────────────────────────┐
   │     │    Collect/collate data    │
   │     └──────────────────────────┘
   │                  │
   │                  ▼
   │     ┌──────────────────────────┐
    ─────│     Review constraints     │
         └──────────────────────────┘
                      │
                      ▼
         ┌──────────────────────────┐
         │      Identify options      │
         └──────────────────────────┘
                      │
                      ▼
         ┌──────────────────────────┐
         │  Select options for detailed │
         │        assessment          │
         └──────────────────────────┘
                      │
                      ▼
         ┌──────────────────────────┐
         │       Assess options       │
         └──────────────────────────┘
                      │
                      ▼
         ┌──────────────────────────┐
         │   Identify preferred option  │
         └──────────────────────────┘
                      │
                      ▼
         ┌──────────────────────────┐
         │   Review preferred option   │
         └──────────────────────────┘
                      │
                      ▼
         ┌──────────────────────────┐
         │      Report assessment     │
         └──────────────────────────┘
```

Table 1 Site and process information requirements

Location information	Process information
National Grid Reference for site, area and location map	Description of process, including flow chart
Distribution of surrounding housing and industry	Land and infrastructure requirements
Electricity grid connections	Input process materials, including fuel, chemicals, raw materials (*e.g.* cooling water)
Availability, source, and preferred transport option of process materials	Output materials, including energy, waste, and other environmental releases
Location of sites of special scientific interest, European Habitat Directive sites, or other environmentally sensitive areas	
Characteristics of local communications and infrastructure	
Relevant economic data related to site, *e.g.* grants, generation tariff, land prices	

engineering, and practical constraints which may limit the choice of location or plant. For example, the decision to enter the mid merit market will restrict the type of plant, and grid pricing structure may pinpoint a particular region as the preferred location. Where decisions which limit the choice of plant or location have been made before undertaking the BPEO assessment, this should be acknowledged together with the reasons for the decision. Other constraints may arise as a result of legislation, regulation, relevant company or national policies, economic and financial considerations, or other technical considerations, and these should be identified and recorded.

Collection and Collation of Data

The collection and collation of data can be divided into two main areas: that relating to potential sites, and secondly, the characteristics of the potential generation processes. The exact information required will depend on the nature of the project being considered. In general, the information required could be considered under the headings indicated in Table 1.

A key decision at this stage is the amount of detail in which the information is collected. The guiding principle should be one of 'fitness for purpose'. For BPEO assessments undertaken to inform the selection of plant and location, it is likely that an assessment based on broad engineering estimates, emission factors, or scoping calculations would be appropriate. Where the assessment is considering a specific process type at a particular location, then more detailed information would be required.

The purpose of this stage is to put together sufficient information so that any further constraints on the selection of site or process can be identified, a wide range of location and process options can be compiled, and a limited number selected for detailed evaluation. Where possible, in order to ease the burden of handling large volumes of data, use should be made of Geographic Information Systems to store and access data, particularly in relation to location. The development of such a system will also enable the data to be readily updated as the study progresses.

Following completion of this stage the data should be reviewed and any further restrictions on the selection of a particular process or location should be identified and this included in the statement of objectives and constraints given in the first stage.

Generation of Options

This is an important step and should be undertaken carefully, since if the 'best' option is not identified at this stage it cannot form the output from the process as a whole. The outcome will be a number of different location/process combinations, each exhibiting a number of unique features. The major variable will be in the characteristics of the process(es) being considered at the location.

The generation of options should be undertaken in a structured manner, for example, the information on site characteristics should be reviewed and local constraints identified. A similar process should then be undertaken for each process option, identifying constraints which are inherent in the process; for example, raw material requirements or pollutant releases and also those which would arise as a result of the implementation of the process at one location rather than another. Options for dealing with the constraints should then be identified. This process should be undertaken for each of the major stages in the process life cycle, *i.e.* construction, operation, and decommissioning.

The options considered should include preventative as well as remedial measures, recycling rather than disposal, and take into account the possibility of transfer of pollutants between environmental media. The techniques applied to generating options can be varied; for example, 'brainstorming' sessions, workshops, or working groups may all be appropriate, depending on the circumstances. Given the wide variation in the nature of constraints, it will be important to include as wide a range of experience as possible in the preparation of the overall list to ensure that viable options are not omitted.

Select Options for Detailed Assessment

It is almost inevitable that a large number of options will be produced by the previous stage and these will need to be screened to identify a manageable number for more detailed analysis. Experience suggests it is difficult to properly evaluate more than about six different location/process options.

The screening exercise should be carried out against defined criteria, one approach might be to rank or score qualitatively each location/process option against the different constraints. Assuming the constraints were ranked or scored from low to high, where a high score represents a significant constraint, options which were persistently ranked or scored at the low end of the scale might be taken through for further assessment. As a minimum, it might be expected that the options selected would be capable of meeting current and likely future legislative and regulatory controls. Inevitably this procedure will involve some value judgements; for example, in assessing the relative importance of many small impacts against a few major ones. Careful thought will need to be given to the presentation of this stage to ensure that the approach, assumptions, and judgements can be clearly understood by a third party. Presentation in the form of a matrix, using colour coded symbols of different size or number to represent the magnitude of an impact, might be helpful.

Option Evaluation

The assessment of options should be undertaken in as quantitative manner as possible and consider both the environment effects and the economic implications of a particular option. In stage 2, data on a wide variety of locations and processes were collected; where possible, this information should form the basis for this assessment. However, it is recognized that more detailed site-specific information may need to be collected at this stage to undertake fully the assessment. The Environment Agency has provided guidance on the types of issues to be included in the Environmental Assessment of projects which impact on the water environment.[5] Whilst the report provides detailed guidance on impacts to the aquatic environment, it does describe a structured approach to the identification of impacts which could usefully be applied to other media.

The impact on the environment of each location and process option combination should be assessed, taking into account cross media transfers, long- or short-term effects and long- or short-range transport through the environment arising from the construction, operational, or decommissioning phases. It is important that options are considered on a consistent basis and that the same methodology is applied across the different options. A major issue in this respect is the definition of the boundary of the option being considered; ideally this should cover impacts arising from all stages in the plant life cycle from production of raw materials to ultimate disposal of waste and decommissioned plant. This should include effects not only of the process in normal operation but also episodic events such as accidental releases.

The RCEP's 12th Report on the BPEO concept provides little guidance on procedures for assessing effects on the environment. Indeed, the requirement to assess effects gives rise to two of the most intransigent problems in environmental assessment, namely: how can the relative magnitude of impacts be compared across different types of effect (*e.g.* toxicity *versus* visual intrusion), spatial scales (global *versus* regional or local), and temporal scales (effects which occur on the scale of decades, such as global warming, *versus* effects which occur over seconds, such as odour); and secondly, how can the relative impacts and costs arising from an option be balanced against each other? This second question will be discussed in more detail in the following stage of the overall assessment process.

Whatever approach is adopted, it should exhibit a number of characteristics:

[5] Environment Agency, *Environmental Assessment: Scoping Handbook for Projects*, HMSO, London, 1996.

(i) Be capable of identifying the option which results in the least environmental impact.
(ii) Be capable of being applied consistently across the range of options.
(iii) Be based on information which is (readily) available.
(iv) Take into account the extent of specialized knowledge required by practioners.
(v) Be practical.

Possible approaches include Life Cycle Assessment (LCA),[6] Risk Analysis,[7] and the Environment Agency's Guidance for Operators and Inspectors for IPC Processes.[8] It is unlikely that a single approach will be appropriate for all aspects of a particular assessment, but elements of these different approaches might be combined to meet the overall needs of the study. For example, LCA[6] makes use of a scoping stage in the assessment to identify impacts which are unlikely to be significant and can therefore be ignored, with commensurate savings in the assessment effort required. Both LCA[6] and Agency Guidance on BPEO assessments for Integrated Pollution Control[8] recognize that impacts cannot be assessed on a single scale; for example, within the Environment Agency's scheme, releases of greenhouse gases are assessed using their global warming potential, whilst the toxicological effects of released chemicals are expressed as a proportion of the relevant environmental standard. Provided the methods are consistently applied, the most appropriate procedure can be selected for assessing the relative impacts of different effects.

A key consideration is how, for a particular option, these disparate measures of effect can be brought together to represent the overall impact of the process. A wide variety of scoring and weighting systems can be devised to combine environmental effects at different scales or of different types. The decision as to which scheme to adopt and how much weight to apply is largely subjective and should be recognized as such in the audit trail.

Alternatively, a cost–benefit approach might be considered in which a monetary valuation of the impacts on people, the environment, and commercial assets is attempted. There is a considerable literature on the economic and social costs of power generation.[9–11] The benefits of such an approach are that it provides a weighting of the different effects on receptors and provides a mechanism for combining measures of disparate effects into a single unit, money. However, there is a very significant degree of uncertainty over the dose response

[6] Society of Environmental Toxicology and Chemistry, *A Conceptual Framework for Life Cycle Impact Assessment*, Society of Environmental Toxicology and Chemistry and SETAC Foundation for Environmental Education, Pensacola, USA, 1993.

[7] Department of Environment, *A Guide to Risk Assessment and Risk Management for Environmental Protection*, HMSO, London, 1995.

[8] Environment Agency, *Best Practical Environmental Option Assessments for Integrated Pollution Control. Volume I: Principles and Methodology. Volume II: Technical Data*, Environment Agency Technical Guidance Note E1, HMSO, London, 1996.

[9] D. W. Pearce, C. Bann and S. Georgiou, *The Social Costs of Fuel Cycles*, HMSO, London, 1992.

[10] M. Holland and J. Berry, *EXTERNE: Externalities of Energy—Volume 1. Summary*, EUR 16520 EN, European Commission, Directorate General XII—Science, Research and Development, Luxembourg, 1995.

[11] *Acid Rain in Europe: Counting the Cost*, ed. H. ApSimon and D. W. Pearce, University College, London, Imperial College, London and Economics for the Environment Consultancy, London, 1996.

functions that underpin the economic analysis, particularly for effects on forestry, water, and biodiversity, and in the assessment and valuation of factors such as visual intrusion, noise, and landscape degradation. Valuation of these factors requires data on people's preferences and their worth, in the context of the project being considered,[11] which may not be available or is expensive to obtain.

The methods adopted therefore need to be properly and clearly justified for third party review. Whilst it is tempting to try and summarize the results of the assessment for a particular option as a single number, it is probably more easily understood if the results are set out in the form of a table which compares the 'score' for each effect arising from each location/process option.

In parallel with the assessment of effects, the capital and operating costs of each location and process option should be estimated and expressed as a net present cost or annualized cost. The basis for the appraisal should be clearly set out to include information on discount rate, treatment of replacement and residual assets, and any allowance for project risk. Guidance on economic appraisal techniques can be found in the HM Treasury publication, *Economic Appraisal in Central Government: A Technical Guide for Government Departments*,[12] and in a Department of Environment, Transport and the Regions (formerly Department of Environment) publication, *Policy Appraisal and the Environment*.[13]

The results of the economic and environmental assessment should be presented in a transparent and consistent manner so that interested parties are able to review and audit the proposals. Outline guidance on the information to be included in the presentation is given by the RCEP[4] and also by the Environment Agency.[8] The presentation should include maps showing the proposed location and key features of interest *e.g.* housing, local infrastructure, and environmentally sensitive areas, *e.g.* Sites of Special Scientific Interest (SSSIs) or European Habitat Directive sites. Process flow diagrams are needed, showing process inputs and outputs for construction, operational, and decommissioning stages in the project, and a description of the assessment methodology adopted, together with the basis for key decisions and assumptions. The environmental assessment may give rise to a variety of quantitative and qualitative information; for the purposes of summarizing the assessment, a pictorial presentation using different size or number of symbols for different environmental factors may be helpful, provided that the basis for assessment of each factor is clearly noted. In the case of cost information, all relevant assumptions should be shown, together with the overall discounted costs of different options. In addition to total values, the cost per unit energy produced may also be helpful.

Identification of Preferred Option

The selection of the BPEO involves a balancing of the environmental effects and economic costs arising from the project and will inevitably be a matter of judgement. The Environment Agency's guidance on BPEO assessments for IPC processes[8] suggests three possible approaches to considering the trade-off

[12] HM Treasury, *Economic Appraisal in Central Government: A Technical Guide for Government Departments*, HMSO, London, 1991.
[13] Department of the Environment, *Policy Appraisal and the Environment*, HMSO, London, 1991.

between costs and environmental effects. Although they are primarily intended for application to industrial processes, the principles can be more widely applied.

Actual cost comparison: annualized or net present costs for each option can be compared with a range of environmental effects. This allows a broad trade-off to be made between options across a range of environmental effects.

Incremental costs compared to incremental environmental effects: the incremental discounted cost can be compared with incremental changes in environmental quality to show the costs of moving to the next less harmful location and process option. This allows the costs of progressively reducing environmental effects to be seen and may illustrate a breakpoint between options which indicate where improvements can only be achieved at greatly increased cost. The approach generally requires a single measure of environmental effect to be set alongside the costs; whilst this may be possible, care should be exercised in the choice of environmental parameter. This is particularly important where effects have been scored on some scale which allows options to be ranked, but the scores do not provide a true reflection of the environmental impact of the effect, *i.e.* the 'true' environmental distance between options is not known.

Incremental costs over and above the option with greatest environmental effects: the discounted cost of each location and process option above the costs of the option with the greatest environmental effect can be compared with changes in environmental effects to show the costs per change in environmental effect. Whilst this allows the magnitude between each option and the option with the greatest environmental effects to be identified, it suffers from similar problems to the previous approach in requiring a single measure of environmental effect to be used in the comparison.

For practical purposes, it might be considered that the BPEO is the 'breakpoint' where the costs of alternative location and process options start to rise considerably compared with the reduction (or improvement) in environmental effects.

Review Preferred Option

Having made a preliminary identification of the BPEO, the assessment should be reviewed to ensure that the decision is robust. It is likely that during the assessment a number of decisions and assumptions will have been made and there will be uncertainties in the data and assessment techniques used, for example in pollutant transport modelling. Therefore, there may well be benefit in undertaking sensitivity analyses. The key question is to determine to what extent data, decisions, or assumptions would need to change in order to arrive at a different conclusion. If the level of change is beyond reason, then the overall BPEO conclusion can be considered as robust.

Report Assessment

The overall assessment should now be brought together for final reporting. It is likely that this will include audiences external to those directly involved in the evaluation and some thought will need to be given to the intended readership. It may be that more than one report is required, bringing out different parts of the assessment process; for example, a wide ranging report may be appropriate for submitting to planning authorities, a report concentrating on pollution control issues may be necessary for environmental protection regulatory authorities, and a summary key issues profile for public information.

In any case, the report(s) should focus on the key decisions, assumptions, and techniques applied, and provide a description and justification of the environmental and economic appraisal methodologies, together with the results of the assessment for each option. The approach used to select the BPEO should be justified and described and the outcome clearly identified, together with the results of any sensitivity analysis. Supporting information should be placed in appendices or separate reports which could be made available on request to regulatory authorities or other interested third parties.

4 Conclusions

The BPEO concept provides a structured approach to the selection of a suitable location and process for power generation. The assessment can be undertaken in two stages: the first to select the location and process type, the second to identify the BPEO in detail for the process at that location.

Key stages in the approach are the definition of the study objectives, data collection, the selection of options for assessment, environmental and economic assessment, selection of the preferred option, and presentation of the results. Running throughout the methodology is the need to maintain an audit trail. The assessment of the environmental and economic implications of the options and the subsequent selection of the BPEO probably represent the greatest challenges in the overall procedure as there is no absolutely objective basis for comparing different environmental effects or balancing their consequences with economic costs. The decision as to which option represents the BPEO therefore needs to be taken on the basis of judgement.

Because of the importance of qualitative judgements in the approach, it is vital that the process is open to external scrutiny and, therefore, the clear presentation of results and the need to maintain an audit trail play a significant part in the procedure.

Environmental Impact of the Nuclear Fuel Cycle

MALCOLM J. JOYCE AND SIMON N. PORT

1 Introduction

The commercial generation of electricity by nuclear means began in the 1950s and has increased steadily to provide 17% of the world's power in 26 countries.[1] There are now over 450 reactors world-wide and further construction continues. Although several types of reactor are in use, the generation principle is similar in all: energy from the fission reaction of ^{235}U nuclei is transferred *via* a coolant to drive turbines which generate electricity.

In common with all methods of electricity generation, the nuclear industry has an environmental impact associated with it, which is the subject of this article. The impact of radiological effects and non-radiological effects are discussed separately and summarized in terms of future implications and global impact. There is a considerable amount of background information required to contextualize technical conventions and established approaches; this is summarized and fully referenced. A glossary of abbreviations and acronyms is included. The article begins with an introduction to the process at the heart of the industry—the nuclear fuel cycle.

The Nuclear Fuel Cycle

Uranium is removed from the ground in an ore from open-pit and underground mines. It is usually present in concentrations of only a few fractions of a percent and is processed and refined into a material known as 'yellowcake' (U_3O_8). This is done at uranium mills, near to the mines, prior to transport to fuel fabrication plants. The solid waste from this activity is known as 'mill tailings'.

At the fuel fabrication plant, yellow cake is further refined to uranium tetrafluoride (UF_4). For some types of reactor,[2] such as PWRs, BWRs, and

---

[1] UNSCEAR, *1993 Report to the General Assembly, with Annexes*, E.88.IX.7, UN, New York, 1993.
[2] D. J. Bennett and J. R. Thomson, *The Elements of Nuclear Power*, Longman, Harlow, 1989.

Issues in Environmental Science and Technology, No. 11
Environmental Impact of Power Generation
© The Royal Society of Chemistry, 1999

AGRs, it is necessary to enrich in ^{235}U. In this case the uranium tetrafluoride will be further processed to uranium hexafluoride (UF_6) and enriched to a few percent. These compounds are then converted to either an oxide or a metal fuel (again dependent on the type of reactor the fuel is to be used in) and machined into fuel rods and assemblies.

The fuel assemblies are loaded into a reactor where thermalized neutrons cause the ^{235}U nuclei to fission and release energy. After several months, the fuel no longer produces energy efficiently. It is regarded as 'spent' and is replaced by new fuel. In addition to ^{235}U and ^{238}U, other nuclides are now present in the spent fuel as a result of the fission of ^{235}U and neutron capture in ^{238}U. These include the *fission products* (the fragments resulting from ^{235}U fission) and the *actinides* (the series of heavier elements that follow actinium in the Periodic Table). The latter include ^{239}Pu, which is fissile like ^{235}U.

Reprocessing is the task of extracting the isotopes of ^{239}Pu and the remaining ^{235}U from the spent fuel so that they can be used as fuel in the future. The alternative to this stage is to store indefinitely or dispose of the complete spent fuel assemblies. There are currently three countries that reprocess on a commercial scale: the UK at Sellafield, France at Cap de la Hague and Marcoule, and Japan at Tokai-Mura.

2 Radiological Impact

Radiation and the Environment

Radiation from the nuclear industry represents a comparatively small, though not inconsequential, component of the total impact of radiation on the environment. When discussing the effect of radiation from the nuclear industry, it is important to consider the other contributions to radiation in the environment. The collective dose commitment for the world population is shown in Figure 1.[1]

The data in Figure 1 assume a 50-year period of continuing practice, for activities such as power generation, medical exposures, *etc.*, and single events from 1945 to 1992 for exposures resulting from weapons tests and severe accidents. The contributions from nuclear power generation (0.28%), severe accidents (0.07%), and occupational exposure (0.07%) are too small to be shown separately and are presented in *Others*. It should be noted that the occupational dose estimate is for all workers exposed to radiation and not just those in the nuclear industry. The percentage contribution of the occupational exposure of workers in the nuclear industry to the world total is 0.014%.

It is clear from Figure 1 that *Natural Sources* represent the largest contribution to radiation dose, and these include:

- Cosmic rays (from space)
- Terrestrial γ-rays (from isotopes in the earth's crust)
- Radionuclides in the body (such as ^{40}K)
- Radon and its decay products (from buildings and the ground)

Whilst there is some distinct variation in background level with location (for

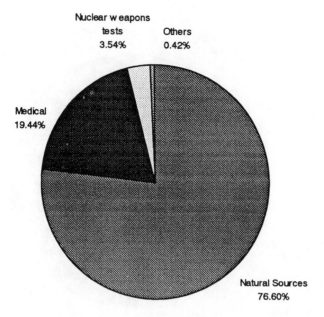

Figure 1 Collective dose committed to the world population

example, the variation in terrestrial γ-ray exposure with geology), the overall level is relatively constant. In the present discussion of the radiological environmental impact of nuclear power, we shall return to the contribution of natural sources for purposes of comparison.

The contribution of the nuclear industry to the world radiation dose is a legacy that distinguishes the environmental impact of nuclear power generation from many other, non-nuclear, generation means. This is an important distinction since ionizing radiation poses a hazard to society through its potential to cause damage to living tissue and consequently to health. The facts that radiation is often invisible and its effects can be long-term further complicate the management and socio-economic understanding of the radiological impact.

The collective dose to the public from nuclear power generation can be subdivided in terms of the various stages of the fuel cycle. This is shown in Figures 2(a) and 2(b).[1] It is necessary to present these data as two charts because the regional impact of nuclear power generation activities is different from the global impact. The latter concerns the effect of solid waste disposal that has a long-term global impact through environmental dispersion.

To understand the radiological impact of nuclear power generation further, it is necessary to introduce several concepts concerning the nature of radiation, its biological effects, and the current approach to quantifying its environmental impact.

Radiation Properties. For the present discussion there are four forms of radiation to consider: alpha (α), beta (β), gamma (γ), and neutron (n). Alpha, beta, and neutron radiation are different types of particle radiation. These particles are ejected by radioactive nuclei. Gamma radiation is a form of very-high-frequency electromagnetic radiation (*i.e.* similar in character to radio waves and microwaves but of a much reduced wavelength). Gamma radiation is also emitted by unstable, radioactive nuclei.

Figure 2 (a) Local and regional component of collective effective dose to the public by stage of the fuel cycle; (b) global component of collective effective dose to the public by stage of the fuel cycle

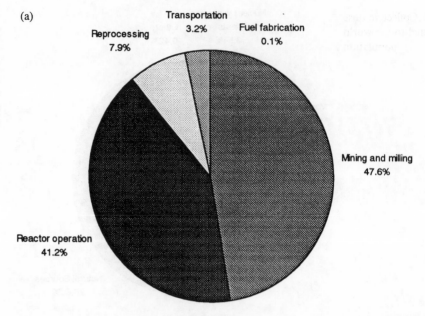

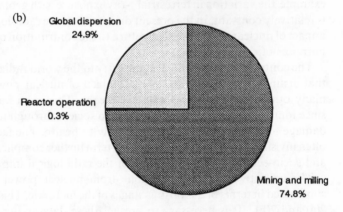

When these forms of radiation interact with surrounding materials they lose their energy *via* ionization. Each form of radiation brings about a characteristic level of ionization. For example, α-particles are intensely ionizing because they are massive, electrostatically charged particles. Consequently, they lose their energy rapidly and do not penetrate very far into absorbing materials—a thin piece of paper will shield from α-particles. In comparison, β-particles (electrons with half the magnitude of the charge of α-particles) penetrate further, although a thin piece of aluminium will stop these. Neutrons and γ-rays are uncharged and are able to penetrate much further. Lead or concrete are often used to shield from these.

The different levels of ionization and penetration of each type of radiation result in different hazards. For example, as an agent external to the body, α-particles pose little threat because our clothes, hair, and the dead, surface skin layer will prevent them damaging the living cells underneath. However, if substances that are α-active are ingested or inhaled, the living tissue of internal organs can be exposed. Neutron, β-, and γ-radiation pose an external hazard because they can penetrate through such protective layers.

Biological Effects. Ionizing radiation can damage living tissue because it breaks up the molecules of the tissue. If the molecules of a living cell are affected, the cell can cease to function properly or to function at all. The effect of radiation is dependent on many variables including radiation type, intensity, energy, and duration of exposure. Biological effects are further complicated because some isotopes are ingested more easily than others, whilst some internal organs accumulate more of a specific isotope than others.

The biological effects of radiation can be considered as either *deterministic* or *stochastic*. Deterministic effects occur in the tissue of the body that is exposed to the radiation—they are *somatic*. Deterministic effects were very prevalent when radioactivity was first discovered and include burns, radiation sickness, and anaemia. As knowledge of these effects has developed, and precautions such as protective clothing have been introduced, such cases now only result from accidental exposures and as possible side-effects of medical radiation treatment.

Stochastic effects are concerned with the statistical aspect of the effects of radioactivity. They can be somatic or hereditary. For example, an inhaled α-emitter might cause damage to a cell such that it mutates and develops into a malignant cancer. This is often a considerable time (decades) after the individual has been exposed, a delay known as the *latency period*. Hereditary effects are those that are manifest in the offspring of the exposed individual. These result from the radiation damage to the germ cells found in the reproductive organs. The stochastic effects of radiation, just like those of any carcinogen, are more difficult to understand, control, and predict than deterministic effects. The International Commission on Radiological Protection (ICRP) recommends safe radiological practice based on advice from key scientific expertise from around the world. The nuclear power generation industry is subject to legal regulation by the independent governmental bodies operating on this advice. These include the Nuclear Installations Inspectorate (NII) of the Health and Safety Executive (HSE) in the UK and the Nuclear Regulatory Commission (NRC) in the USA.

Quantifying the Environmental Impact of Nuclear Power

To assess the environmental impact of nuclear power, it is necessary to quantify the effects of radiation. Three types of radiation dose estimate are important:

- The maximum dose to individuals and specific populations known as *critical groups*
- The mean dose to an individual averaged over an exposed population
- The total dose commitment to a local, regional, or global population

In this section the conventions and principles behind radiation exposure and the associated dose estimates above are introduced.

Radiation Sources. The atomic nuclei of some isotopes are unstable and will decay spontaneously to become another nuclear species. These exist naturally all around us and even in our bodies. This process is accompanied by the emission of radiation. The rate at which atomic nuclei undergo this process is measured in terms of the SI unit the *bequerel* (Bq), which is equal to one disintegration per second.

Each radionuclide has a characteristic rate of decay which is defined by the time it takes for half the number of nuclei in a substance to decay. This is known as the *half-life* ($t_{1/2}$) and is measured in seconds (s), minutes (m), days (d), and years (a). This provides a measure of the activity of the radionuclide.

Radiation Dose. Radiation imparts energy to the material it interacts with. A measure of this energy per unit mass of the absorbing material is known as the *absorbed dose* and is measured in *grays* (Gy). One gray is equal to one joule per kilogram ($J\,kg^{-1}$).

The different degree of ionization brought about by each type of radiation is reflected by the introduction of another parameter called the *equivalent dose*. This parameter is equal to the absorbed dose multiplied by a radiation weighting factor[3] which reflects the level of ionization of the specific radiation concerned. To reflect the variation in response of the different tissues and organs in the body, a tissue weighting factor is also introduced.[3] The separate effects can then be summed over the human body to give the *effective dose*. Although the weighting factors are dimensionless, equivalent dose and effective dose are given the units of *sieverts* (Sv) to distinguish them from absorbed dose.

The effective dose provides a measure of the stochastic detriment to an individual as a result of exposure. Recommended limits of exposure are provided by the ICRP.[3] For situations where the exposure of a specific selection of the population is of interest, an average individual dose can be multiplied by the number in the population. In several estimates given later, a model is used that assumes population density of $25\,km^{-2}$ out to a distance 2000 km from a nuclear plant, *i.e.* a total population of approximately 3×10^8. This is used in conjunction with the relevant atmospheric and meteorological models of activity dispersion. The units for this *collective effective dose* are *man-sieverts* (man Sv).

Dose relating to power generation is often expressed as a function of the amount of electricity-equivalent produced *i.e.* man Sv per gigawatt-year [man Sv $(GW\,a)^{-1}$]. This takes into account variations in plant in terms of electricity production and provides a normalization of the environmental impact independent of changes in global generation capacity.

Dose Commitment. A further development of effective dose is required because many environmental effects of radiation exposure will occur long into the future. For example, radioactive species in the environment are responsible for a degree of exposure now but also over the time they are present. This is taken into account by the *dose commitment* which is the integral of the average effective dose over a period of time for a specific population, again measured in sieverts (Sv). For long-lived isotopes the time period used is often infinity. Studies of radionuclide

---

[3] *Ann. ICRP*, 1990, **21**, 1–3.

transport and uptake by the biosphere[4–6] are used to reflect the population affected by the isotope concerned. Hence, for long-lived, readily transported isotopes, such that they are eventually dispersed globally, it is necessary to evaluate several estimates of dose commitment. These would indicate the local/regional commitment and global commitment.

Environmental Impact

The radiological environmental impact of nuclear power generation can be divided between the effects of:

- Radioactive waste:
  - Aerial effluents
  - Liquid effluents
  - Solid waste
- Accidents:
  - Chernobyl, USSR (1986)
  - Three Mile Island, USA (1979)
  - Windscale, UK (1957)

In this section we shall discuss the radiological environmental impact of nuclear power generation following the plan above.

Aerial Effluents. Small amounts of effluent are routinely discharged from a number of sources throughout the nuclear fuel cycle. In the UK, it is recommended that the annual individual dose from man-made sources should not exceed 1 mSv (excluding medical treatments). This compares with 2.2 mSv per year from natural sources and includes a recommended limit of 0.5 mSv per year from effluents of the nuclear industry.[7]

At the uranium mining/milling stage, of greatest environmental concern is the release of radon. The natural radioactive decay of uranium results in a number of radioactive daughter products as shown below:

$$^{238}U \xrightarrow{\alpha} {}^{234}Th \xrightarrow{\beta^-} {}^{234m}Pa \xrightarrow{\beta^-} {}^{234}U \xrightarrow{\alpha} {}^{230}Th \xrightarrow{\alpha} {}^{226}Ra \xrightarrow{\alpha} {}^{222}Rn$$

| 4.47×10^9 a | 24.1d | 1.17m | 2.45×10^5 a | 7.54×10^4 a | 1600a | 3.82d |

The radon isotope ^{222}Rn is a hazard *via* inhalation because it is a gas and a short-lived α-emitter. It can be released from uranium mines as an airborne discharge, or from mill tailings that are stored in piles. The latter are often covered with a solid layer or water barrier to reduce the release of radon. The local/regional collective effective dose has been estimated using a reference mine

[4] G. Desmet, P. Nassimbeni and M. Belli (eds.), *Transfer of Radionuclides in Natural and Semi-natural Environments*, Elsevier Applied Science, London, 1990.

[5] M. W. Carter, J. H. Harley, G. Schmidt and G. Silini (eds.), *Radionuclides in the Food Chain*, Springer, Berlin, 1988.

[6] P. J. Coughtrey (ed.), *Ecological Aspects of Radionuclide Release*, Blackwell Scientific, Oxford, 1983.

[7] *Review of Radioactive Waste Management Policy*, Cmnd paper 2919, HMSO, London, 1995.

and surrounding population densities together with an atmospheric dispersion model.[6] This gives an estimate of 1.5 man Sv $(GW a)^{-1}$, mostly by inhalation. There is considerable variation between sites.

Because the half-lives of ^{230}Th and ^{226}Ra are long, radon will continue to be emitted for a long time. Consequently the global environmental impact is heavily dependent on the way in which mill tailings are dealt with in the future. Uncovered, the average exhalation rate of radon from mill tailings has been estimated[8] to be 20 Bq $m^{-2}s^{-1}$. It is worthwhile to compare this to the natural background exhalation rate of normal soils, which is 0.02 Bq $m^{-2}s^{-1}$. The treatment of mill tailings in the future will result in rates between these two extremes.

The important radionuclide emissions from the fuel fabrication stage are the few heavy neighbouring isotopes of uranium: ^{234}U, ^{235}U, ^{238}U, ^{234}Th, and ^{230}Th. Since ^{230}Th is long-lived, this isotope acts as a barrier to the creation of isotopes further down the decay chain. The main exposure mechanism for fuel fabrication is inhalation because most of the isotopes listed above are α-emitters. Using the population model described earlier, the collective effective dose from aerial effluents of fuel fabrication has been estimated as 0.0028 man Sv $(GW a)^{-1}$.

Power plants discharge gases as a result of purging and degassing of coolant and other routine treatment operations. These gases arise *via* the corrosion of activated parts of the reactor structure or the leakage of fission products from the fuel through damaged fuel-cladding. The following gases are typical:

- Noble gases (^{133}Xe, ^{135}Xe most abundant)
- Tritium and tritiated compounds
- ^{14}C (in carbon dioxide)
- Halogens (specifically isotopes of iodine)

The amounts discharged vary considerably, dependent on fuel type, reactor design, and effluent treatment procedures in use. The emission of all effluents is regulated by the NII (UK) or NRC (USA) in accordance with the recommendations of the ICRP on limits of aerial discharge.[3]

Finely divided particulates may also be discharged with the aerial effluent, of which ^{137}Cs and ^{60}Co contribute most significantly to the dose. Particles of fuel and other actinides have also been reported.[9] The normalized emissions of aerial effluents over the period 1980–89 are shown in Figure 3.[1] World-wide, the collective effective dose for reactor effluents over 1970–1989 is shown in Figure 4,[1] which also includes liquid effluents.

Several approaches are adopted to treat aerial effluent prior to emission, including high-efficiency particulate (HEPA) filters, activated charcoal, liquid scrubbing, and catalytic conversion. The potential dose can be reduced by storing the effluent for a short time to allow the short-lived isotopes to decay before they are vented to the atmosphere.

Measurements of the emissions are taken to ensure that emitted doses fall within ICRP recommendations. The immense dilution by the atmosphere reduces the levels of radionuclide in the environment to such an extent that they are too small to be measurable except at the point of emission. The various

[8] UNSCEAR, *1988 Report to the General Assembly, with Annexes*, E.88.IX.7, UN, New York, 1988.
[9] A. C. Chamberlain, *Sci. Total Environ.*, 1996, **177**, 259–280.

Figure 3 Normalized activity [GBq (GWa)$^{-1}$] *versus* calendar year for aerial emissions from reactors

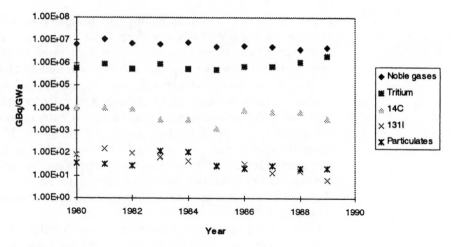

Figure 4 Collective dose (man Sv) for effluents released from reactors worldwide

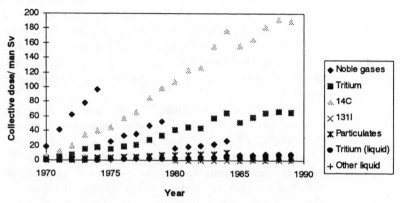

transport processes of the radionuclides in the environment are then modelled to provide an estimate of public exposure.[8,10] These measurements estimate the local/regional effective dose for all reactor operation over the period 1985–1989 to be 1.4 man Sv (GW a)$^{-1}$,[3] which includes liquid effluents.

Reprocessing plants have a high absolute throughput of radionuclides but, because the throughput of fuel is small, the dose per electricity-equivalent of fuel reprocessed can appear low. The isotopes of concern in aerial effluent from reprocessing plants are ^{3}H, ^{14}C, ^{85}Kr, and ^{129}I and particulates. Aerial effluents are treated to remove ^{129}I and undergo caustic scrubbing and filtration before being released. The normalized dose as a function of the energy-equivalent of the fuel reprocessed for local/regional commitments is 0.05 man Sv (GW a)$^{-1}$.

The composition of the effluent streams from reprocessing plants varies according to plant design and treatment procedures, as is the case for the mining and power generation stages of the nuclear fuel cycle. However, there are the additional variables of the type of fuel reprocessed and the time for which it has been left to cool. These aspects are managed by the reprocessing plant such that aerial effluent emissions fall within regulatory guidelines.

[10] R.J. Pentreath, *Nuclear Power, Man and the Environment*, Taylor & Francis, London, 1980.

Liquid Effluents. For mining/milling activities in dry areas, liquid effluent is not an issue. In wet areas, where surface water can act to carry away spoil, in particular ^{226}Ra, effluent streams are treated prior to their release to the environment. Doses due to liquid effluent from mining activities are small in comparison with aerial effluents. This is also true for fuel fabrication.

Many of the isotopes in liquid effluents from reactors and reprocessing have short half-lives that enable them to be dealt with *via* storage as described for aerial effluents. The isotopes of most environmental interest are the longer-lived, more abundant species, such as ^{3}H, ^{14}C, ^{129}I, ^{134}Cs, ^{137}Cs, and the actinides. The chemistry of tritium (^{3}H) is identical to hydrogen and, as such, it can replace the hydrogen atoms in water molecules comprising the aqueous/organic medium of the liquid effluent stream. The local/regional normalized collective effective dose estimate is 0.2 man Sv (GW a)$^{-1}$.

Critical groups are exposed *via* the ingestion of local produce (often seafood as many nuclear sites are near the coast and discharge to the sea), external whole-body irradiation from the local environment (often beaches and coastal areas because physical mechanisms, such as sea-spray transport, result in higher activities here), and localized external β-doses to the fishermen handling fishing equipment. The levels of exposure for these groups peaked in the mid-1980s and have declined since to 0.1–0.3 mSv for Sellafield, with other sites having similar estimates.

The ingestion of radioactivity from liquid discharges, *via* beaches, fish, and seafood, has been linked with a significant excess of childhood leukaemia cases in Seascale, near Sellafield, over the period 1968–78. An independent study[11] has identified a correlation between the excess incidence of leukaemia and high recorded occupational radiation exposures of the fathers of the children affected. The fathers had worked at Sellafield and the link is therefore considered occupational and not due to the liquid discharges. This link has been independently discussed[12] and refuted[13] in the light of current knowledge of hereditary causes of leukaemia.

Solid Radioactive Waste. The nuclear power generation industry is unusual in that its waste products have been under considerable regulatory scrutiny long before environmental awareness became commonplace across industry as a whole. Indeed, for many organizations *waste management* forms a major part of the nuclear industry's business. This business often extends to managing radioactive waste from other, unrelated, sources such as medical and research institutions.

Uranium ore extraction. The important solid wastes from this stage are the mill tailings. The main environmental concern here is the exhalation of radon gas, as discussed earlier.

Fuel manufacture. A consequence of enrichment is that a great deal of the less-fissile isotope ^{238}U results as a waste product known as *depleted uranium*. Whilst a very long-lived α-emitter, it has few requisite qualities except its high

[11] M.J. Gardner, *Int. Stat. Rev.*, 1993, **61**, 231–244.
[12] M.F. Greaves, *Leukaemia*, 1990, **4**, 391–396.
[13] R. Doll, H.J. Evans and S.C. Darby, *Nature*, 1994, **367**, 678.

density. It can be used in *fast-breeder reactors*[2] to manufacture ^{239}Pu for use as fuel (this is not currently widespread since it is expensive). Depleted uranium has found limited application as counterweights in aircraft and in military armour and a great deal of it remains stockpiled at fuel fabrication plants in the form of uranium hexafluoride. Across the USA there is an estimated 560 000 tonnes of this gas stored in cylinders. The future uses of depleted uranium hexafluoride are currently under discussion.[14] A possible application is in the shielding of spent-fuel casks.[15] The refining of uranium hexafluoride would also liberate fluorine, which has many industrial uses.

Other than depleted uranium, fuel manufacture produces solid waste with similar radiolytic properties to that produced by mining and milling, but in much reduced amounts. This is dealt with as low/intermediate level waste as discussed below.

Power generation/reprocessing. The fission process produces waste in the form of fission products and actinides, as discussed earlier. The fission products are a diverse mix of many radioactive isotopes with a broad range in half-life, often β/γ-emitters. In comparison, the actinides are long-lived and are predominantly α-emitters (there is often associated γ-ray emission too).

Solid wastes also arise from solidified materials used in treatment procedures (discarded ion-exchange resins, *etc.*). Further waste is produced *via* the interactions of neutrons with the surrounding materials of the reactor. This is an important aspect of decommissioning operations of old nuclear plant.

Dealing with Radioactive Waste

The choice of whether to store waste indefinitely or dispose directly to the environment depends on the category of the waste concerned, of which there are four:[16]

- Very-low-level waste (VLLW)
- Low-level waste (LLW)
- Intermediate-level waste (ILW)
- High-level waste (HLW)

Each category is defined in terms of the specific activity of the waste. Qualitatively, these guidelines define VLLW as being suitable for disposal with ordinary refuse, the majority of this being non-nuclear. LLW is defined as having activities that are small enough to allow disposal to the environment by existing routes, dependent on the level and nature of the LLW activity. ILW is defined as containing considerable activity, but not generating heat, which requires treatment before direct disposal. HLW is heat generating and must be isolated from the environment. Treatment methods of each level of waste are described

---

[14] *Programmatic Environmental Impact Statement for Alternative Strategies for the Long-term Management and Use of Depleted UF$_6$*, DOE/EIS-0269, 1998.

[15] P. A. Lessing, *The Development of 'DUCRETE'*, INEL-94/0029, 1995.

[16] *Radioactive Waste Management*, Cmnd 8607, HMSO, London, 1982.

Figure 5 Volume (m³)
against year for waste
stocks and arisings in the
UK

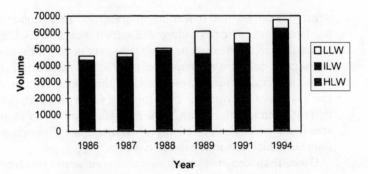

below. Figure 5 shows the volume of waste stocks and arisings for sites in the UK over the period 1986–94.[17]

Very Low-level Waste. The upper limits for VLLW are 400 kBq per 0.1 m³ β/γ, with no one item exceeding 40 kBq β/γ.

Low-level Waste. Low-level waste is composed of any slightly contaminated rubbish from any establishment that uses radioactive material. Whilst this represents a considerable volume of debris, it is comparatively low when compared with the millions of tonnes of toxic waste generated throughout the world, and much of it is reduced by incineration/compaction prior to disposal. The upper limits for LLW are 4 GBq tonne⁻¹ α-activity, 12 GBq tonne⁻¹ β-activity. Two approaches have been adopted for disposing of LLW: shallow trench storage and deep sea disposal.

Shallow trench storage. The ground is a source of natural background radiation, in the form of terrestrial γ-rays (with annual effective dose levels of several mSv in some areas[1]). Hence, waste of low levels can be buried with little radiological detriment to this environment. The LLW is buried in trenches that are typically[18–20] 20 m deep by 20 m wide by a few hundred metres long. Trenches often have a lining of impermeable material to prevent direct water movement and they are covered by at least one metre of compacted soil. The trenches drain into a tidal stream from which any low-level activity is diluted to insignificance at sea. There are many such sites containing material from many uses besides nuclear power production. In the UK, LLW is stored at BNFL Drigg, near Sellafield in Cumbria, and at UKAEA Dounreay, Caithness.

Deep sea disposal. This practice ceased following a temporary moratorium in 1982 which was converted to a permanent ban in 1994, subject to review every 25 years.[21] LLW of activities greater than those ideal for shallow trench disposal was disposed of in the deep ocean, beyond the continental shelf, at a depth of

[17] Data from: *The UK National Environmental Health Action Plan* 1996, DOE/DH HMSO, London 1996.
[18] A. V. Pinner and M. D. Hill, *Radiological Protection Aspects of Shallow Land Burial of PWR Operating Wastes*, NRPB-R138, 1983.
[19] NRC-US, *Licensing Requirements for Land Disposal of Radioactive Waste*, NUREG-0945, 1982.
[20] US Geological Survey Circular 1036, *Proc. of the LLW Disposal Workshop, Big Bear Lake, Calif.*, ed. M. S. Bedinger and P. R. Stevens, US Dept. of the Interior, Washington, 1987.
[21] *This Common Inheritance: UK Annual Report*, Cmnd 2822, HMSO, London, 1995.

several kilometres.[22] Some ILW was also disposed of in this way. This operation was controlled by the Convention on the Prevention of Marine Pollution by the Dumping of Wastes within limits drawn up by the International Atomic Energy Agency (IAEA). These limits are set according to independent radio-oceanographic research that assesses the possibility of LLW being concentrated by a food-chain, *etc*. The disposal process itself was controlled by the Nuclear Energy Agency (NEA) of the OECD and each disposal required a specific licence issued by the Ministry of Agriculture, Fisheries and Food (MAFF) for UK disposals. The contaminant of concern was usually tritium. The waste was packaged in large, concrete-lined drums. These were designed to withstand the impact with the sea-bed and to remain intact for at least the time it takes for the radioactivity to decay away.

To put this practice into perspective, it is instructive to consider briefly the natural radioactivity of the sea. The world's oceans are naturally radioactive because of massive amounts of several isotopes; the major α- and β-activities are due to ^{238}U and ^{40}K, respectively. The activity of the world's oceans can be estimated[23] to be, on average, $390\,Bq\,m^{-3}$ for ^{238}U and $130\,kBq\,m^{-3}$ for ^{40}K. If it is assumed that the waste disposed of at sea does not find a route back to man as a concentrated form, the low-level waste will be diluted to insignificance at well below these background levels.

Some dumping was carried out in the USA prior to regulation in areas that are now considered too shallow.[24] Independent studies of these dumps[25] indicate low detrimental environmental effect.

Intermediate Level Waste. Intermediate level waste accounts for the largest volume fraction of waste, estimated as $70\,000\,m^3$ by 2000 AD in the UK alone.[16] It can be subdivided into two separate categories according to half-life. The longer-lived α-active species, originating from fuel pin assemblies, pond sludges, and plutonium reprocessing contaminants, will be radioactive for thousands of years. These materials are isolated from the environment by immobilizing them chemically, with concrete or bitumen, and they are then stored temporarily, under controlled conditions, above ground. Whilst this is a satisfactory solution in the short term, anxieties about the suitability of this approach over thousands of years have stimulated several geological studies in Germany, Switzerland, the USA (Yucca Mountain, Arizona[26]) and the UK (an investigation led by UK NIREX). The goal of these scientific assessments is to determine whether the longer-lived intermediate level waste can be stored safely in a *deep repository*. At the time of writing, the geological studies continue concerning groundwater flow, gas migration, and chemical containment. There is not an immediate need for the repository since the waste is expected to be stored above ground for at least 50 years.

[22] NEA, *Review of the Continued Suitability of the Dumping Site for Radioactive Waste in the North-East Atlantic*, OECD-NEA, 1985.

[23] J. M. Prospero and F. F. Koczy, *Encyclopedia of Oceanography*, Reinhold, New York, 1966, vol. 1.

[24] J. B. Lewis, *Atom*, 1983, **317**, 49–52.

[25] P. J. Taylor, *The Impact of Nuclear Waste Disposals*, Political Energy Research Group, Oxford, Res. Rep. 8, 1982.

[26] *Phys. World*, 1998, **11**, 11.

The second category of intermediate-level waste comprises short-lived β/γ-emitting nuclides containing few long-lived α-emitters. This waste is made up of irradiated reactor components and will decay to very low levels in a few hundred years. Hence, it can be stored in a *shallow repository* (20 m deep), similar to the shallow trenches used for LLW. Such a repository has water-tight trenches to reduce contamination spreading with water flow.

High-level Waste (HLW). Ninety-five percent of the total waste *radioactivity* from nuclear power generation is in the form of HLW. This is refined from spent fuel during reprocessing operations. The decision whether to reprocess the waste strongly influences the amounts of LLW/ILW produced and resides with the commercial judgement of the fuel owner. HLW contains the fission fragment and actinide streams. Whilst it is the most radioactive of the waste categories, it represents the smallest volume, as shown in Figure 5. The radioactivity in HLW generates considerable heat and it requires cooling during storage.

There are two distinct stages in dealing with HLW. The HLW is extracted from the spent fuel such that the fission-fragment/actinide mixture is an aqueous stream. This is then stored in large, double-skinned, stainless-steel tanks that are shielded with large amounts of concrete. These tanks are cooled by coils that circulate coolant through them.

The HLW tanks are temporary holding vessels for HLW since waste in the solid phase is safer and easier to manage. To convert the HLW waste to the solid phase, a process called *vitrification* has been adopted. Studies have shown[27] that, of all solids, glasses are particularly resistant to corrosion and leaching over long periods of time. Vitrification involves encasing the waste in a glass matrix. The vitrified HLW is contained in large stainless-steel churns that are stored above ground whilst the shorter lifetime isotopes decay away (around 50 years). After this time, much of the heat generation will have reduced too. The vitrified HLW is cooled by circulating air around the vessels. This has the added benefit that the air can be monitored to assist in the management of the waste.

One option for the disposal of vitrified HLW is that it will be placed in a deep repository, similar to that under discussion for ILW. However, there are additional hazards to be considered. The process of site selection had not started at the time of writing.

Accidents

Accidents in the nuclear industry can vary from minor non-radiological anomalies, like those encountered in many other industries, to major accidents of radiological significance. The International Nuclear Event Scale (INES) enables prompt communication of the safety significance of such events and is shown in Figure 6.[28] Events rated level 2 and above are defined as having some radiological significance. Table 1 provides data for incidents at UK sites, over the period 1984–1993, in terms of their frequency and INES rating.[29] Very few

---

[27] M. J. Plodinec, *J. Non-cryst. Solids*, 1986, **84**, 206–214 and further to 299.

[28] From the Nuclear Utilities Chairman's Group (NUCG).

[29] *The Prospects for Nuclear Power in the UK*, Cmnd 2860, HMSO, London, 1995.

Figure 6 The INES ratings

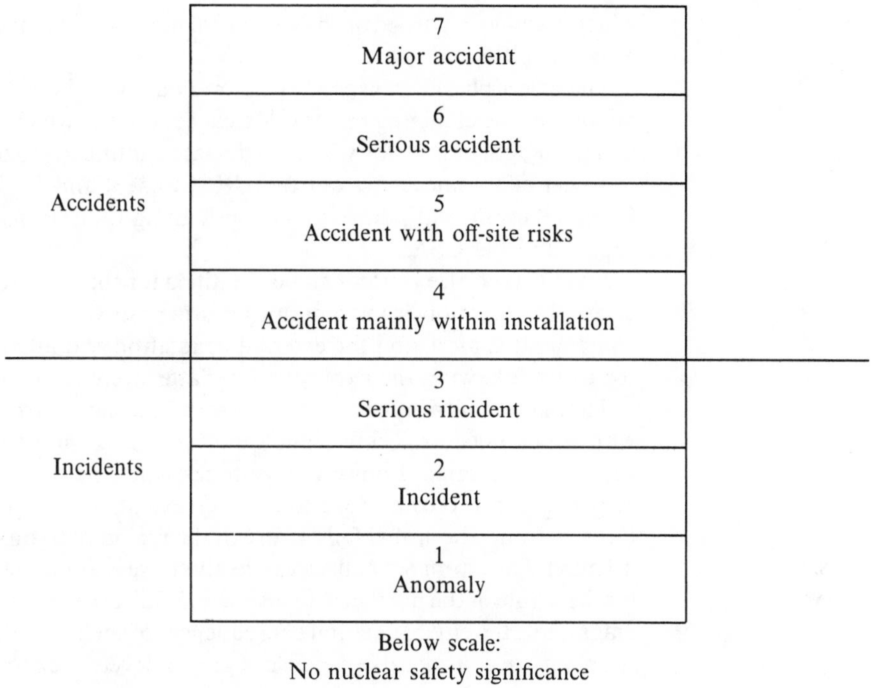

	7 Major accident
Accidents	6 Serious accident
	5 Accident with off-site risks
	4 Accident mainly within installation
Incidents	3 Serious incident
	2 Incident
	1 Anomaly

Below scale:
No nuclear safety significance

Table 1 Data for incidents at UK sites over 1984–93[29]

Year/Level	0	1	2	3	> 3
1984	425	6	1	0	0
1985	393	16	1	0	0
1986	407	23	2	0	0
1987	420	17	2	0	0
1988	368	12	1	0	0
1989	386	35	2	0	0
1990	268	27	1	0	0
1991	176	78	2	0	0
1992	215	58	0	1	0
1993	180	52	4	0	0

incidents have featured at level 2 with only one at level 3. The contrasting large number of level 0 incidents reflects the thoroughness of the reporting system.

In this section we shall discuss three severe accidents in terms of their environmental impact: Chernobyl (rated level 7), Three Mile Island (rated level 5), and Windscale—an event strictly of military origin but still of considerable radiological significance and of relevance to this article. Whilst these events have specific relevance to the current environmental impact of nuclear power, they also indicate the environmental implications of nuclear accidents in general.

Chernobyl, USSR, 1986. The Chernobyl accident occurred on 26 April 1986 in the Number 4 reactor of the Chernobyl power station, north of Kiev in the USSR. It involved a explosion which released large amounts of radioactivity to the environment. There are many references describing the accident in detail,[1,2,8] which is widely regarded as the worst civil nuclear incident to date, resulting in 30 direct fatalities.

The atmospheric transport of the radioactivity from Chernobyl is complex because of the changing meteorological conditions which followed the accident and the length of time for which the release continued (10 days). Most of Europe was eventually affected, from the UK, Sweden, and Finland in the north, to Greece, Kuwait, and Turkey in the south. Long-range transport reached Canada, Japan, and the USA.

Several exposure routes can be identified for the Chernobyl fallout, including external irradiation from radionuclides deposited on the ground, ingestion of contaminated food, and the external irradiation and inhalation of activity from the cloud following the explosion (the latter were short term).

The major contributors to the dose from Chernobyl are ^{131}I, ^{134}Cs, and ^{137}Cs with other nuclides being significant for a short-lived external γ-dose from deposited material. Long-lived isotopes, such as ^{3}H, ^{14}C, ^{85}Kr, and ^{129}I, contribute to the total dose to a lesser extent. The highest effective doses in Europe, from Chernobyl fallout, are in the region of 0.6 mSv (*cf.* natural sources 2.4 mSv). The estimated collective effective dose is approximately 600 000 man Sv for the whole of the northern hemisphere. Whilst these exposures are not of great magnitude, the environmental consequences of such an accident in the vicinity of the plant are considerable because of greater levels of exposure, mass population resettlement, and on-going decontamination. Long-term effects, such as the increased incidence of congenital deformity and thyroid cancer, the subject of much investigation throughout the world[30–37]

Following the Chernobyl accident, a network was set up between several nations to enable prompt action in response to nuclear incidents that affect several countries.[38]

Three Mile Island, USA, 1979.   The Three Mile Island (TMI) accident happened on 28 March 1979, and involved the partial destruction of the Unit 2 PWR reactor core owing to a loss-of-coolant incident.[2] Whilst containment was held successfully, a release of radioactivity to the atmosphere occurred and coolant water leaked, so contaminating an auxiliary building.

The major release was an aerial escape of noble gases, specifically the isotopes of ^{133}Xe, ^{133m}Xe, and ^{135}Xe, estimated as 370 PBq in total.* There was also a release of ^{131}I estimated as 1 TBq in total.[39] A wide sampling programme detected trace levels of radioactivity (of the order 0.01 Bq) in the environment. In particular, ^{3}H

*^{133m}Xe is a long-lived (of the order of days) isomeric state of the ^{133}Xe nuclide.

[30]   J. Icso and M. Szollosova, *Radiat. Prot. Dosim.*, 1998, **77**, 129.

[31]   E. Williams, *J. Br. Nucl. Enery Soc.*, 1997, **36**, 443.

[32]   E. E. Buglora, J. E. Kenigsberg and N. V. Sergeev, *Health Phys.*, 1996, **71**, 45.

[33]   D. L. Henshaw, *Br. Med.J.*, 1996, **312**, 1052.

[34]   S. Yamaslita and S. Nagataki, *Thyroid*, 1998, **5**, 153.

[35]   E. D. Williams, F. Pacini and A. Pinchera, *J. Endocrinol. Invest.*, 1995, **18**, 144.

[36]   J. Royasburke, *J. Nucl. Med.*, 1992, **33**, N23.

[37]   C. Groner, *New Sci.*, 1998, **160**(2155), 20.

[38]   *The Radioactive Incident Monitoring Network*, HMSO, London, 1993.

[39]   T. M. Gerusky, in *Radionuclides in the Food Chain*, ed. M. W. Carter *et al.*, Springer, Berlin, 1988, p. 157.

Table 2 Estimated collective dose to populations 0 to 50 miles from TMI over the period March 28–April 15 1979[39]

Radius/ miles	Population	Collective dose/ man Sv	Average dose/ mSv
0.4–1.0	324	0.19	0.586
1–2	1 816	0.36	0.198
2–3	7 579	1.20	0.158
3–5	18 567	1.80	0.097
5–10	137 474	7.20	0.052
10–20	577 288	11.73	0.020
20–50	1 420 071	5.37	0.004
Total	2 163 579	27.85	0.013

was detected in nearby water, ^{131}I was detected in milk, and ^{137}Cs was found in the soil. However, such measured levels are below background radiation levels. The most significant radiation exposure was the external β/γ component from the noble gas plume. Ground-level dose rates from this cloud are difficult to assess because of the dispersion/dilution mechanisms and complex meteorological influence.

Dose estimates for the accident were taken from the study of TLDs located around the plant. The average effective dose was estimated as 0.1 mSv, whilst the maximum was estimated as 1 mSv. These estimates are well below background and recommended occupational exposure limits.[3] The estimated collective doses to the surrounding populations, up to a distance of 50 miles, are given in Table 2.[39] For comparison, the total collective dose for the population within this area from natural sources is estimated at 2400 man Sv.[40]

The release from TMI is considered so small that no detectable increase in radiation-induced health effects are expected. However, many studies of the environmental effects of TMI continue, including the TMI Mother–Child Registry and the TMI Cancer Study. These report every five years and have found no measurable hereditary or prenatal effects as a result of TMI. The TMI Cancer Study has not reported any enhanced cancer incidence as a result of the accident. Nevertheless, because of the long latency period of many possible associated effects, these early studies are widely regarded as premature. Recent reports,[41,42] in which the topographic and meteorological aspects of the plume transport have been included, disagree[43] as to whether the incidence of cancer has increased in the vicinity of TMI as a result of the accident. A small, short-term increase has been identified[44] in the first three years after the accident, which is attributed to the effects of stress or increased health awareness.[45] The severity of the environmental impact is the subject of considerable discussion and research[46–48]

[40] J. I. Fabrikant, *Health Phys.*, 1981, **40**, 151–161.
[41] M. C. Hatch, J. Beyea, J. W. Nieves and M. Susser, *Am. J. Epidemiol.*, 1990, **132**, 397–412.
[42] S. Wing, D. Richardson, D. Armstrong and D. Crawford Brown, *Environ. Health Perspect.*, 1997, **105**, 52–57.
[43] M. Hatch, J. Beyea and M Susser, *Environ. Health Perspect.*, 1997, **105**, 12.
[44] M. Hatch, S. Wallenstein, J. Beyea, J. W. Nieves and M. Susser, *Am. J. Public Health*, 1991, **81**, 719–724.
[45] P. S. Houts, G. K. Tokuhata, J. Bratz, M. J. Bartholomew and K. W. Sheffer, *Am. J. Public Health*, 1991, **81**, 384–386.
[46] B. Molholt, *Proc. Workshop on TMI Dosimetry II*, 1985, A109–A111.
[47] B. Molholt, *Proc. Workshop on TMI Dosimetry II*, 1985, C99–C100.

The clean-up operation at TMI involves taking the damaged fuel out of the core, disposing of 2.1 million gallons of contaminated water, and decontaminating the auxiliary building. This task is an expensive, on-going waste disposal problem.[49,50]

Windscale, UK, 1957. The Windscale accident happened on the 10 October 1957 at what is now the Sellafield nuclear site, Cumbria, UK. It is described in detail in several sources.[51-53] The accident resulted in the aerial release of a considerable amount of radionuclides following a fire in the core of a reactor used for the manufacture of plutonium, polonium, and tritium. Data are available on the deposition of radionuclides around the plant[9,54] and also on that which occurred as the cloud travelled firstly north-east and then south-east.[55]

The main concern of the Windscale accident was the ^{131}I contamination of local pastures used for milk production.[56] Iodine is concentrated through cows in their milk and hence, if ingested, the ^{131}I would collect in the thyroid and pose a hazard, even though the contamination of the pasture itself was considerably lower. Samples were taken at a nearby village, Seascale, showing levels that were unacceptable for consumption by children. Distribution was halted for milk with a ^{131}I content of $3.7\,kBq\,l^{-1}$. All deliveries from herds within an area of approximately $520\,km^2$ of Windscale were banned for 20 days (longer in some areas closer to Windscale). Waste milk was tipped into the sewers to be diluted to insignificance at sea.

The contribution of ^{210}Po to the collective dose, *via* inhalation, was significant, as was that of ^{137}Cs in the long term *via* external irradiation from ground deposits. The maximum individual dose (to the thyroid of a child in the Windscale area) is estimated[57] as 160 mSv. The collective effective dose equivalent to the UK population is estimated[57] as 1.9×10^3 man Sv (annual collective dose from natural sources is 1×10^5 man Sv).

A significant cluster of Down Syndrome cases in Dundalk, Ireland, have been linked with the Windscale accident.[58] This has been discounted[59] both in terms of the direction in which the cloud moved and radiolytic knowledge of the syndrome.

3 Non-radiological Environmental Impact

Transportation

Materials used in the nuclear fuel cycle are transported by road, sea, air, and rail.

[48] T. Seo, *Proc. Workshop on TMI Dosimetry I*, 1985, 237–252.

[49] W. Booth, *Science*, 1987, **238**, 1342.

[50] S. Shulman, *Nature*, 1989, **338**, 190.

[51] *Accident at Windscale, No. 1 Pile on October 10th, 1957*, Cmnd 302, HMSO, London, 1958.

[52] *Final Report of the Technical Evaluation Committee*, Cmnd 471, HMSO, London, 1958.

[53] L. M. Arnold, *Windscale 1957. Anatomy of a Nuclear Accident*, Macmillan, London, 1992.

[54] A. C. Chamberlain and H. J. Dunster, *Nature*, 1958, **182**, 629–630.

[55] N. G. Stewart and R. N. Crooks, *Nature*, 1958, **182**, 627–628.

[56] K. F. Baverstock and J. Vennart, *Health Phys.*, 1976, **30**, 339–344.

[57] M. J. Crick and G. S. Linsley, *Int. J. Radiat. Biol.*, 1984, **46**, 479–506.

[58] D. Black, *Br. Med. J.*, 1987, **294**, 591–592.

[59] P. M. Sharp and D. J. McConnell, *Br. Med. J.*, 1984, **289**, 378.

Stringent radiological precautions, such as compensated flask design and double-hulled vessels, ensure that the risk of catastrophic environmental effects are minimized. Similarly, transport of mill products, yellowcake, UF_6, and UO_2, contained in suitable transportation containers and subject to current regulation, are unlikely to result in any environmental effects.

In comparison with other fuels, such as coal, oil, and gas, very little uranium is needed to produce comparable supplies of electricity. Hence, the transport requirement of uranium and its compounds throughout the fuel cycle is much less than for conventional fossil fuels. This results in a reduced environmental impact due to air pollution *via* exhaust gases. Although the volume of radioactive material being transported is growing, and will continue to grow in those nations where the nuclear industry grows, in comparison with the enormous transportation requirements for fossil fuels the amount is forecast to remain small.

Mining Waste and Effluent

The initial waste from mining consists of waste rock from the excavation process, similar to any mining procedure. Provided this is free from contamination, such as pyritic material, it is typically employed in the construction industry *via* earthworks, foundations, and roads. Excess rock is normally deposited near to the mine. Liquid effluent from uranium mining consists chiefly of surface water runoff and from ore stockpiles *via* water seepage through the waste rock. This effluent may contain dissolved minerals and suspended solids, which will include uranium and its decay products. Typically, treatment procedures involve pond settling of solids and surface evaporation, use as process feed (water) to the uranium mill, controlled dilution, and discharge during heavy rainfall after the removal of radium-226.

Milling

Tailings slurry is the major chemical waste, apart from radiological waste, from the milling processes. The stream consists of a slurry of leached solid ore and waste solutions from the grinding, uranium purification, leaching, washing, and precipitation processes within the mill. The particle size of the tailings, the type of materials (clay or slimes), porosity, and permeability affect the tailings' ability to retain water. Thus, dewatering techniques are employed to consolidate the mass and return water to the plant, hence reducing the environmental impact. Typical acid leach processes contain sulfate ions, along with soluble metal ions and traces of organic solvent. Sodium removal is achieved using carbonate leaching processes, but this also produces a sulfate waste solution.

Solid tailings are neutralized and consist chiefly of unleached rock and precipitated mineral hydroxides. The tailing slurry system is designed to retain all solids with the liquid effluent and, depending on the climate, is either retained or concentrated *via* evaporation. Even with specifically designed impoundment areas, seepage and runoff can take place, with the risk of affecting aquatic environments. However, the level of pollutants actually leaving *via* the milling process may well be reduced considerably by sorption processes within the tailings themselves.

The environmental impact associated with the mining and milling process can be summarized as arising from mine water, mining waste rock, overburden from open pits, tailings slurry, iron and aluminium hydroxide sludges, gypsum sludges, mill residues, neutralization mill effluents, gaseous effluents, organic compounds (*i.e.* flotation agents, solvents), and biological elements such as algae and fungi. On occasion it is necessary to treat waste waters from the processes to minimize the environmental impact of the organic content, pH, suspended solids, toxic materials, colour, and volatile components. This will normally employ processes such as sedimentation, coagulation, neutralization, sand filtration, ion exchange, precipitation, and biological treatments.

Hexafluoride Process Effluents

Depending on the process used for hexafluoride production, the effluent from the wet and dry processes differ significantly. The majority of the impurities entering from the yellowcake are removed in the wet process raffinate solution *via* solvent extraction. For the dry process, most of the impurities waste from the fluorination and distillation stage are contained in the solid. The wet process effluents consist of:

- Small amounts of CaF_2 from the fluorination step
- Caustic effluents along with residue fumes from the recovery of HNO_3, HF, and the general treatment of off-gas streams
- Neutralization of aqueous raffinate from the solvent extraction process

As for the milling waste treatment, settling ponds are employed after neutralization, with final evaporation leading to burial of the sludge or transfer to a retention system. The raffinate stream accounts for about $5 \, m^3 \, t^{-1}$ of the uranium processed and contains substantially dissolved solids, radium, and thorium-230 entering from the yellowcake feed, and about 0.2% of the uranium processed. Disposal of this effluent is a major problem associated with the wet processing.[60] In addition to this, some scrubber effluents are treated with lime, thus precipitating fluoride ions during settlement in ponds. Typically these are disposed of as CaF_2 *via* burial. The alternative to wet processing—dry processing—provides a non-volatile ash containing iron, calcium, magnesium, copper, and other fluorides, amounting to typically 0.1 tonne per tonne of UF_6 produced.[61]

Effluents from Enrichment Processes

Enrichment plants typically generate only small quantities of fluorides, nitrogen

[60] US Energy Research and Development Administration, *Final Environmental Impact Statement, Expansion of US Uranium Enrichment Capacity*, ERDA-1543, 1996.
[61] M. B. Sears, R. E. Blanco, B. C. Finney, G. S. Hill and R. E, Moore, *Correlation of Radioactive Waste Treatment Costs and the Environmental Impact of Waste Effluent in the Nuclear Fuel Cycle—Conversion of Yellowcake to UF$_6$ Pt. 1, The Fluorination—Fractionation Process*, ORNL-NU-REG/TM7, 1977.

oxides, and sulfur as airborne emissions through the process cooling system, clean-up operations, and associated production facilities. Low levels of sulfates, chlorides, fluorides, and nitrates, plus sodium, calcium, chromium, and iron ions are also observed. Effluents are discharged and diluted owing to their very low levels. It should be noted that by far the largest environmental impact related to the enrichment process is the emission of particulates, nitrogen oxides, and sulfur from the generation of electrical energy through fossil fuel combustion traditionally used to power this process.

Effluents from Fuel Fabrication and Manufacture

The major environmental impact from the fuel fabrication and manufacture stage is non-radiological in nature.[62] For fuel fabrication, HF is the major effluent of airborne interest. The fluorine used for UF_6 manufacture becomes a waste product of the enriched UO_2 powder. The gas streams are scrubbed and filtered to remove this fluoride waste. The liquid effluents, from fuel fabrication, normally contain nitrogen-based compounds from UO_2 powder production and from the nitric acid used in the recovery processes. The air scrubber water is combined with the liquid waste to precipitate CaF_2 *via* the addition of lime. The calcium fluoride formed is filtered and stored on site. The use of fluid bed hydrolysis reduction techniques has substantially reduced gaseous, liquid, and solid effluents in recent years. However, it should be noted that the non-radiological environmental impact of fuel fabrication is minor in comparison to that of the mining and milling processes.

Overall, the major environmental impact is due to the emissions arising from electricity used during the enrichment phase—typically these are formed from fossil fuel electricity production. Therefore, the use of nuclear energy for electricity production and the knock-on effect for the fuel cycle will see the decrease of this environmental problem. The mining environmental impact is no greater than the problems encountered at most metal mines and coal mines. Mining regions vary significantly, but typically commercial-grade uranium ore contains heavy metals which can be leached to the environment and lead to environmental degradation.[63]

Reactor Operation

As described above, during normal operation, nuclear reactors discharge aerial effluents within regulatory guidelines. These effluents comprise a range of radionuclides and consequently a selection of elements. However, since the amounts discharged are small, the non-radiological impact is low. In comparison to emissions from fossil fuel power generation, the discharges of a nuclear reactor are negligibly small.

The major environmental impact from reactor operation arises from thermal pollution. Like many methods of generating electricity, nuclear reactors produce

[62] USAEC Fuels & Materials Directorate of Licensing, *Environmental Survey of the Uranium Fuel Cycle*, WASH-1248, 1974.

[63] D. R. Davy (ed.), *Rum Jungle Environmental Studies*, AAEC/E365, 1975.

waste heat. This reaches the environment through the discharged coolant stream. This is typically greater with a nuclear reactor compared to a fossil fuel power station. A nuclear station rejects all its heat to the cooling system water, whereas a fossil fuel plant rejects only 15% *via* its cooling stacks along with the combustion products.[64] Thus a nuclear station will emit 50% more heat to the waters surrounding it than a fossil fuel station producing a similar amount of electricity. There have been reports that there is very little effect[65] and, conversely, that a small rise in temperature can stimulate the growth of parasitic organisms and fungi that can lead to bacterial fish disease.[66] It is clear that the effect of thermal pollution on the environment depends on the dilution characteristic of the polluted medium.

4 Future Implications and Summary

Future

The expansion of the nuclear power industry in the UK, the USA, and Sweden has declined owing to the competing economic viability of other generation means, principally gas,[29] and public opposition as a result of anxieties concerning the environmental impact of nuclear power. However, several nations, such as France, Japan, and India, currently have reactors under construction. For the developing world, where rapid growth in living standards necessitates increased power generation, nuclear power is an attractive option, especially for those nations with little natural fossil fuel resource of their own.

The disparity between the costs of nuclear power and fossil fuels in the UK, where gas is considerably cheaper, has stimulated claims that the cost of energy production from fossil fuels does not include the cost of the associated environmental detriment. However, it can be argued that the cost of decommissioning old nuclear plant is also outside of the unit energy cost of nuclear power generation.

In comparison with fossil fuels, nuclear power stations produce very little carbon dioxide which is suspected to be responsible for global warming. This is also true for sulfur dioxide (SO_2) and oxides of nitrogen (NO_x) that are responsible for acid rain. It is estimated that 70% SO_2/NO_x comes from fossil fuel energy generation.[29] To meet the emissions requirements for these gases, set under the *Sustained Development* initiative, nuclear power offers a potentially attractive source of energy. However, this not only presumes a precise link between carbon dioxide and global warming but also ignores other competing approaches, such as improved energy efficiency. Indeed, nuclear power is an inflexible source in terms of the long build time (~ 10 years) of the plants, large power outputs (1–2 GW), and long service life of the stations.

[64] G. M. Masters, *Introduction to Environmental Science and Technology*, Wiley, Chichester, 1975.

[65] R. F. Pocock, *Nuclear Power*, The Institute of Nuclear Engineers, 1977.

[66] R. Curtis and E. Hogan, *Perils of the Peaceful Atom*, Gallancz, London, 1970.

Global Impact

Radionuclides with long half-lives, such as ^{3}H, ^{14}C, ^{85}Kr, and ^{129}I, will be in existence long enough for them to achieve global dispersion. Predicting the impact of this scenario is complex because little is known about future population size and world demography, long-term global meteorology, or radionuclide transport mechanisms over timescales of 10 000 years and beyond. In respect of these uncertainties, the most reliable collective effective dose information is bounded at 10 000 years. Beyond this time the uncertainty in dose estimates becomes too great to be of any realistic use. Over 10 000 years the complete global impact of the present period of nuclear power generation is estimated as 123 000 man Sv.[1] This corresponds to ~ 0.028 mSv per individual in the Northern Hemisphere and is almost completely due to ^{14}C.

The global environmental impact of nuclear power is dependent on waste management policies of the future. Currently, because storage is required whilst radionuclides decay to suitable levels, temporary measures such as interim storage are acceptable. In the long term, different measures will be required to ensure that the risk to the environment is minimized. Such measures are currently the subject of much political discussion and scientific research.

In the future, an aspect of considerable importance will be the decommissioning of old nuclear plant. This is an issue that has yet to be undertaken by the world at large. The current intention for old reactors is to demolish the surrounding buildings after a period of 5–10 years. After a period of 100 years, when the remaining radioactivity will have decayed away to a large extent, the core, *etc.*, would be demolished and treated as ILW. The technical challenges of this type of operation are currently the subject of much research.

In summary, the environmental impact of nuclear power generation is significant but not disproportionate in comparison with other electricity generation means. As the world's desire for energy continues to grow, it is perceivable that the nuclear industry will expand, especially in the developing world. As to whether the environmental impact grows in proportion is dependent on the strict adherence to safety, as the key objective, and the informed regulation of waste and industrial discharges. Such measures will effectively combat the major potential environmental disbenefits of the nuclear fuel cycle.

5   Glossary

AGR – Advanced Gas-cooled Reactor
BWR – Boiling Water Reactor
HLW – High-Level Waste
HSE – Health and Safety Executive
IAEA – International Atomic Energy Agency
ICRP – International Commission on Radiological Protection
ILW – Intermediate-Level Waste
INES – International Nuclear Event Scale
LLW – Low-Level Waste
MAFF – Ministry of Agriculture, Fisheries, and Food

NEA   – Nuclear Energy Agency
NII   – Nuclear Installations Inspectorate
NRC   – Nuclear Regulatory Commission
NRPB – National Radiological Protection Board
OECD – Organization for Economic Co-operation and Development
PWR   – Pressurized Water Reactor
TLD   – Thermal Luminescent Dosemeter
VLLW – Very-Low-Level Waste

Electric and Magnetic Fields and Ecology

DAVID JEFFERS

1 Introduction

The electric and magnetic fields from power lines fall into the 'extremely low frequency' (ELF) portion of the non-ionizing radiation spectrum, which is shown in Figure 1. In 1996, the European Commission[1] produced a report on 'Non-ionizing radiation, sources, exposure and health effects'. The document was produced by a team drawn from the radiation protection authorities in Italy, Germany, and the United Kingdom and included amongst its members the Chairman, Vice-Chairman, and Scientific Secretary of the International Commission on Non-Ionizing Radiation Protection (ICNIRP), which is the formally recognized non-governmental organization in non-ionizing radiation for the World Health Organization. The conclusions open with the statement that 'The most significant source of exposure to non-ionizing radiation for the general public is the sun'. We are concerned here with the ecological effects of power line fields, but it is worth noting at the outset that the energy from the sun, which is essential to life on earth, arrives as electromagnetic radiation and exposure to electric and magnetic fields is ubiquitous. Quite apart from the radiation from the sun, the earth has its own natural magnetic and electric fields which can be comparable in magnitude to those produced by overhead wires.

As the report points out, solar ultraviolet radiation has well-established adverse health effects including skin cancer, but these risks appear to be acceptable to many people. In contrast, there has been increasing public concern about possible, but not established, risks of exposure to extremely low frequency electric and magnetic fields such as those generated by power lines and other electrical equipment.

This introduction has referred to 'non-ionizing radiation', 'electromagnetic fields', and 'electric and magnetic fields'. The term radiation implies that the frequency is high enough for radio or optical waves to be propagated. In

[1] European Commission, *Public Health and Safety at Work. Non-ionizing Radiation, Sources, Exposure and Health Effects*, Office for Official Publications of the European Communities, Luxembourg, 1996, ch. 8, p. 157.

Issues in Environmental Science and Technology, No. 11
Environmental Impact of Power Generation
© The Royal Society of Chemistry, 1999

Figure 1 The non-ionizing radiation spectrum

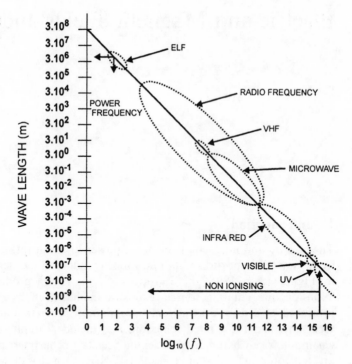

Figure 2 Electric fields (E) and magnetic fields (H) in a plane propagating wave

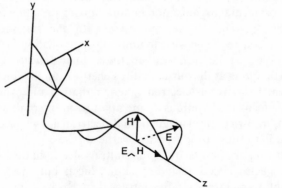

non-ionizing radiation, the photon energy is too low to disrupt a covalent bond. This requires 12.4 eV, and puts the boundary between ionizing and non-ionizing radiation in the ultraviolet band.

Figure 2 shows the electric and magnetic fields which make up a plane propagating wave; they are at right angles to each other and in a fixed ratio.

$$\frac{\text{Electric field } (\text{V m}^{-1})}{\text{Magnetic field } (\text{A m}^{-1})} = \frac{E}{H} = 377 \ (\text{ohm}) \qquad (1)$$

This relationship applies to a plane wave in the 'far field', well removed from its source, but the power density S (W m^{-2}) is given by the general result

$$S = E \times H \qquad (2)$$

where S, the cross product of the electric and magnetic field vectors, is known as Poynting's vector.

After this brief discussion of non-ionizing radiation and electromagnetic fields, it has to be noted that, in spite of its title, the International Commission on Non-ionizing Radiation Protection also concerns itself with frequencies which are too low to allow the propagation of radiation to any significant degree, and ELF fields fall into this category.

The wavelength of radiation is given by

$$\text{Wavelength (m)} \times \text{Frequency (Hz)} = \text{Velocity of light (m s}^{-1}) = 3 \times 10^8 \, \text{m s}^{-1}$$

or

$$\lambda f = c \tag{3}$$

The 50 Hz of the electricity supply thus has a wavelength of 6000 km, which is far longer than any overhead line to be found anywhere in the world. Because the lines are very short compared to the wavelength, they do not, for practical purposes, propagate radiation. As a consequence, their electrical and magnetic fields can be considered in isolation and are not in fixed ratio to each other.

2 Electric Fields at the Conductor Surface

There is a common misconception that electric power flows in the conductors of overhead lines and underground cables. In fact, as Poynting showed in 1884,[2] the power transfer takes place in the electric and magnetic fields surrounding the conductor and not in the conductor itself. The power density is given by Poynting's eponymous vector (shown in eqn. 2) and electric and magnetic fields are not some unwanted by-product of the transfer of electric power: they are the medium in which the power transfer takes place. Poynting produced his analysis five years before the installation, in 1889, of the country's first recognizably modern style high voltage power station at Deptford. In the subsequent 100 and more years, electricity has become essential to modern life, but Poynting's vector still describes its transfer. However, it must be admitted that most engineers obtain the 'correct' answers to their design calculations by assuming that the power does flow in the cable.

A very large number of papers has been published on the biological effects of electric and magnetic fields. Most of them have concentrated on the magnetic field, but when the environmental effects of overhead lines are being considered it is more appropriate to start with the electric field because its presence can be readily perceived without the need for instrumentation. In particular, the acoustic noise generated by some overhead lines in damp weather is a consequence of the electric field at the surface of the conductors.

3 Electric Fields and their Effects

Discharges at the Conductor Surface

Figure 3 shows a typical 400 kV line which consists of two three-phase circuits,

[2] L. Solymar, *Lectures on Electromagnetic Theory*, Oxford University Press, New York, 1984, ch. 5, p. 134.

Figure 3 Double circuit, 400 kV, line with two conductors per phase

one on either side of the tower (pylon). Each phase of the three-phase circuit consists of a bundle of conductors suspended from the insulators. There are two conductors in the bundle on the photograph, but heavy duty 400 kV lines have four conductors bundled together. Single conductors are used on some 275 kV lines and on 132 kV lines. In a three-phase circuit, the voltages on the three phases reach their maxima in sequence with an interval of one third of a cycle, *i.e.* 6.67 milliseconds on the UK 50 Hz system. In electrical terms, they are separated in phase by 120°.

The 400 kV is the root mean square (rms) voltage between any pair of conductor bundles in the circuit and the peak voltage is $\sqrt{2}$ (1.414) times this value. The voltage between a conductor and the ground is the phase-to-phase voltage divided by $\sqrt{3}$ (1.732) or 230 kV (rms), 326 kV (peak), for a 400 kV line. The National Grid network has a design span length of 360 m for its 400 kV lines and the minimum height above ground in open country is 7.6 m. Lines only come down to this level, however, when they are at their maximum design temperature, which is 50–90 °C, depending on type. High voltage networks incorporate a high

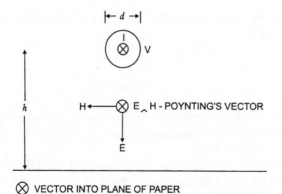

Figure 4 Fields around a single conductor

H - POYNTING'S VECTOR

⊗ VECTOR INTO PLANE OF PAPER

degree of redundancy, so that lines can be switched out without interrupting the supply to consumers; as a consequence, they are normally loaded to no more than 30% of their maximum carrying capacity. Temperatures are normally well below their maximum design values and lines do not 'sag down' to their minimum allowable levels. In practice, minimum heights above ground will be around 9 m.

In the simple analysis which follows, the electric fields around the single conductor shown in Figure 4 are estimated and the results are used to obtain the orders of magnitude of the fields to be expected around the line and their environmental importance.

Electric fields are generated by electric charges, and we are thus concerned here with the net charge on the overhead line conductors. The electric field at ground level is at a maximum beneath the conductors and, in this position, it is dominated by the charge on the lowest conductor. We can, therefore, obtain good estimates of the magnitudes of the fields by considering a single conductor in the lowest position. No attempt will be made to incorporate the empirical correction factors which account for surface roughness, but details of the design process may be found in the *Transmission Line Reference Book*,[3] which is written from an American perspective, or in *Modern Power Station Practice*,[4] which considers British designs.

The charge per unit length, q ($C\,m^{-1}$), on the conductor is given by:

$$q = CV \tag{4}$$

where C ($F\,m^{-1}$) is the capacitance per unit length and V is the voltage relative to ground.

A single wire, diameter d (m), at height h (m) above ground has a capacitance relative to ground of

$$C = \frac{2\pi\varepsilon_0}{\ln\left(\dfrac{4h}{d}\right)} \tag{5}$$

where ε_0 is the permittivity of free space, $8.85 \times 10^{-12}\ F\,m^{-1}$.

[3] Electric Power Research Institute, *Transmission Line Reference Book*, EPRI, Palo Alto, 1987, ch. 3, p. 63 *et seq.*

[4] British Electricity International, *Modern Power Station Practice*, Pergamon, Oxford, 1991, vol. K, ch. 3, p. 64 *et seq.*

The application of Gauss' law gives the electric field at the surface of the conductor as

$$E_1 = \frac{2V}{d \ln\left(\dfrac{4h}{d}\right)} \qquad (6)$$

This most important parameter is the starting point for our environmental discussion because the generation of acoustic noise, ions, ozone, and oxides of nitrogen all depend on it.

Consider now the values for the commonest 400 kV conductor on the UK network which has a diameter of 28.6 mm, a minimum design height of 7.6 m above ground, and voltage relative to ground of 230 kV (rms). Equation (6) shows that the surface field would be 2.3 million volts per metre ($MV\,m^{-1}$) rms or 3.3 $MV\,m^{-1}$ peak. The electrical breakdown strength of air is 3 $MV\,m^{-1}$ and if a single conductor were used in the way we have considered here the electrical discharge activity at its surface would make it unacceptably noisy.

In order to minimize noise production, single conductors are therefore not used at 400 kV. The ones considered in this example are typically used either in pairs spaced 305 mm apart or in quadruples at the corners of a similar sized square. This has the effect of reducing the surface field by increasing the effective diameter of the conductor bundle and sharing the charge between the members of the bundle. As a consequence, the surface electric stress is reduced to the order of 1.6 $MV\,m^{-1}$ for the twin conductors and 1 $MV\,m^{-1}$ for the quadruples.

The electrical discharge activity which can often be heard in damp weather is known as corona and is the artificial equivalent of the natural phenomenon of 'St. Elmo's Fire', which can sometimes be observed in the rigging of ships and at the tips of trees in mountainous regions when the natural atmospheric field is high in thundery weather. Corona noise is irritating, but as we have just seen, conductor bundles are sized to avoid its onset in 'normal' conditions. However, protuberances on the wires like water droplets and wind blown debris such as pieces of vegetation and dead insects give rise to local enhancements of the electric field which can lead to corona discharges. This electrical activity can give rise to ozone, oxides of nitrogen, and air ions which have all been associated with biological activity. As a consequence, it is an obvious question to ask if they can be formed around high voltage wires and then migrate to ground level.

In the UK, overhead line voltages have been limited to 400 kV, but higher levels are in use in other countries. Ozone production was considered when 765 kV lines were introduced on the American Electric Power (AEP) system in the USA. Long term monitoring in 1970–71[5] at 20 locations along the line failed to detect any significant incremental ozone concentrations at ground level.

Similar measurements were made in the vicinity of a 750 kV line that was in heavy corona during foul weather. Ozone was measured 9 m and 0.6 m above ground, but it was only detected at the 9 m position and then only in foul weather. Measured 1-h incremental values were in the range 0–8 ppb. Indoor tests on the

[5] Electric Power Research Institute, *Transmission Line Reference Book*, EPRI, Palo Alto, 1987, ch. 4, p. 202.

production of oxides of nitrogen due to corona show that its production rate is about one quarter that of ozone.

Air ions are formed around high voltage conductors during corona discharges. The discharge is triggered by a photon breaking a molecule of nitrogen or oxygen into an electron and a positive ion, and, when the conductor is at negative potential, the electron is accelerated away from the surface and creates further ions by collision with air molecules. As a consequence, electrons sweep forward, creating positive ions in their wake, and an electron may make about 2×10^5 collisions before it is captured to form a negative ion. Air ions are created naturally in the environment by background radioactivity and cosmic radiation. The end products are relatively stable, forming two groups of ions with narrow mobility ranges: small ions of average mobility $1.5 \times 10^{-4} \, \text{m}^2 \, \text{V}^{-1} \, \text{s}^{-1}$ and large ions of average mobility $3 \times 10^{-8} \, \text{m}^2 \, \text{V}^{-1} \, \text{s}^{-1}$. The ambient densities are, however, subject to large and frequent fluctuations.

Measurements were made over a period of one year of ion concentrations beneath the conductors of a 400 kV transmission line at Bramley[6] near Basingstoke. For much of the time, values under the line showed no difference from background levels. Peaks above background were occasionally witnessed during corona discharge and concentrations exceeded the expected ambient variation of 50–1000 cm^{-3} for approximately 2% of the time during the year's survey.

4   Electric Fields at Ground Level

We have seen in the previous section that the electric fields on the surface of 400 kV conductors are in excess of one million volts per metre. The field at ground level in our single conductor example is:

$$E_0 = \frac{2V}{h \ln\left(\dfrac{4h}{d}\right)} \tag{7}$$

and, with the values chosen, the field E_0 is thus 8.7 kV m^{-1}. However, as we saw above, the electric field on the surface of the conductor is unacceptably high and bundle conductors have to be used. For a bundle of n conductors of diameter d (m) on a pitch circle diameter of A (m), the equivalent diameter of the bundle is

$$D = (ndA^{n-1})^a \tag{8}$$

where $a = 1/n$, and the field at ground level is

$$E_0 = \frac{2V}{h \ln\left(\dfrac{4h}{D}\right)} \tag{9}$$

The conductors of a 'quad' bundle are spaced 305 mm apart, giving a pitch circle diameter of 431 mm and an equivalent diameter, D, of 309 mm. With this diameter, the ground level field is 13.2 kV m^{-1}. This illustrates a typical

[6] R. Houlgate, in *Atmospheric Ions and Industrial Activity*, The Institution of Electrical Engineers, London, 1986.

D. Jeffers

Figure 5 Electric field at 1 m above ground due to double circuit, 400 kV, line

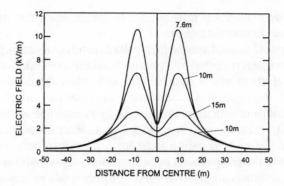

compromise of engineering design: bundles of conductors are used to limit the electric field at the conductor surface, but their use has the effect of increasing the field at ground level from $8.7\,\text{kV}\,\text{m}^{-1}$ to $13.2\,\text{kV}\,\text{m}^{-1}$. The calculations we have done for illustration here slightly overestimate the field to be expected beneath a transmission line because the field generated by the other conductors of the circuit partially cancel that due to the bottom one. Figure 5 shows the electric fields to be expected around a transmission line running at 400 kV with quad conductor bundles (type L6 on the UK system) and the maximum field to be expected is of the order of $11\,\text{kV}\,\text{m}^{-1}$. On a 400 m long span of such a 400 kV line, the conductor 'sag' is around 15 m and, at the tower supports, the lowest conductor is more than 20 m in the air. With this clearance the electric fields are reduced to less than $2\,\text{kV}\,\text{m}^{-1}$. The effect of ground clearance on the electric field is shown in the figure.

Electric field lines start and finish on electric charges and, when a conducting object is placed in an electric field, the field causes a displacement of electric charge. As a consequence, if a person (say) stands below a transmission line, there will be a negative electric charge induced on his body surface when the charge on the conductor above his head is positive and *vice versa*. These induced charges can make the hair on the skin vibrate and this, in turn, can be perceived as a tingling sensation. Tests carried out on 136 volunteers in the USA[7] gave a median value of $7\,\text{kV}\,\text{m}^{-1}$ (at 60 Hz) for the perception level on the hand.

Because the field induced on the body surface is alternating, it will generate currents inside the body and a person wearing leather soled shoes, which give good electrical contact with the ground, will have a current of about $15\,\mu\text{A}$ per $\text{kV}\,\text{m}^{-1}$ flowing through his shoes into the ground. This current of 0.15 mA in a $10\,\text{kV}\,\text{m}^{-1}$ field is much less than the threshold of about 1 mA from hand to hand for the perception of steady currents. It can, however, be sufficient to make a fluorescent tube glow faintly when held up below a power line. This demonstration is beloved by objectors to overhead lines, but it is worth noting that it is also possible to make a tube flicker with the natural static charges induced by shuffling ones feet on a nylon carpet. For best effect, the free end of the tube should be touching grounded metalwork, such as a central heating radiator.

The charges which are induced in conducting bodies give rise to enhanced

[7] Electric Power Research Institute, *Transmission Line Reference Book*, EPRI, Palo Alto, 1987, ch. 8, p. 365 *et seq.*

fields at their extremities. The field at the top of a person's head, for example, is about 18 times the unperturbed value. If trees are allowed to grow beneath high voltage wires, the fields induced at the tips of branches may be large enough to give rise to corona which damages leaves and stunts growth. In the limit, vegetation could become close enough for an arc to be struck from the live wire and cause the protective system to operate. To avoid this happening, trees must be cut back in the vicinity of an overhead line. In the UK[8] a clearance of 3.1 m is specified to trees which cannot support a ladder, and in orchards a clearance of 5.3 m is required. In some countries, the electrical utility establishes a 'right of way' in which the line is sited and development is controlled. The Bonneville Power Administration, for example, which operates a high voltage network in the north west of the USA, has a right of way width of 32–50 m for its 500 kV lines. In England and Wales, the electrical network has been developed without the use of such rights of way. Towers and lines have 'way leaves' which allow passage over a land owner's property, and whilst access is granted for maintenance and necessary tree cutting, the responsibility for land management remains with the land owner.

The induced currents which were considered above are the 'direct' effects of electric fields. There are also important indirect effects which give rise to 'micro-shocks' when a person or animal touches a charged object. Work by the Bonneville Power Administration established that the electric fields caused by transmission lines affect the health and mortality of honey bees inside wooden hives and the most likely explanation is that the bees receive small but frequent shocks. These adverse effects can be mitigated by screening the hive with wire mesh. Animals can experience similar shocks if their drinking troughs or fences become charged, but these effects can be minimized by earthing the offending objects.

As far as humans are concerned, the most common sources of micro-shocks are vehicles parked beneath overhead lines, open umbrellas, and vegetation which grows in spikes. In dry weather, the body of a vehicle can become insulated from ground by its rubber tyres and a person wearing leather shoes can discharge the vehicle to ground when he touches it. Tests by the Electric Power Research Institute[9] in the USA found that the short circuit currents in such situations ranged from $0.39\,\text{mA}/(\text{kV m}^{-1})$ for a school bus at 60 Hz to $0.06\,\text{mA}/(\text{kV m}^{-1})$ for a farm tractor, *i.e.* $0.32\,\text{mA}/(\text{kV m}^{-1})$ and $0.05\,\text{mA}/(\text{kV m}^{-1}1)$ at 50 Hz. One can readily see that touching a large vehicle, such as the bus, can give shock if it is parked in an unperturbed field of (say) $5\,\text{kV m}^{-1}$.

An umbrella presents a large area to the field for the collection of charge, and it can be discharged by touching its metal shaft. Tests by the Electric Power Research Institute obtained a median value of $2\,\text{kV m}^{-1}$ for the threshold of perception for people using umbrellas.

If one is wearing insulating shoes and vegetation which grows to sharp points brushes against ones legs, shocks can be experienced in high electric fields. This effect can be a problem in high voltage substations if weeds are not controlled.

[8] Electricity Association, *Overhead Line Clearances*, technical specification 43-8, Electricity Association, London, 1988.

[9] Electric Power Research Institute, *Transmission Line Reference Book*, EPRI, Palo Alto, 1987, ch. 8, p. 357.

A hypothesis has been advanced that the presence of an electrical field can enhance the deposition, on the body, of the radioactive decay products of naturally occurring radon.[10] Such deposition undoubtedly takes place on energized wires in the laboratory, but recent measurements failed to detect any enhancement beneath a 400 kV line at Didcot.[11]

Electric fields are readily distorted by conducting objects and it is these distortions which give rise to the field enhancements discussed above. It should also be noted, however, that one also finds field reductions around such objects and a tree, for example, provides ground level screening in its vicinity. The inside of a house is largely (90% or more) shielded from external electric fields.

5 Magnetic Fields

Magnetic fields, at the levels to be expected below overhead lines, cannot be perceived without the aid of instruments. Strictly speaking, the unit of the magnetic field is the ampere per metre $(A\,m^{-1})$ and that is the unit used when calculating Poynting's vector. However, most biological and epidemiological researchers give their data in terms of the magnetic flux density, B, for which the unit is the tesla. In air and biological tissue the two are related by

$$\mu_0 \times \text{ magnetic field } (H/A\,m^{-1}) = \text{ magnetic flux density } (B/T), \text{ where } \mu_0 = 4\pi \times 10^{-7}$$
(11)

B and H are both loosely referred to as 'the magnetic field'.

The tesla (T) is an enormous unit, and so environmental fields are normally quoted in microtesla (μT) and $1\,A\,m^{-1} = 0.8\,\mu T$. As with the electric field, we can obtain the expected order of magnitude of the magnetic flux density below an overhead line by assuming that its value is dominated by the current in the lowest conductor.

For a straight wire carrying a current I (amp), the magnetic field at distance h (m) is given by

$$H = \frac{I}{2\pi h}$$
(12)

and the flux density by

$$B = \frac{\mu_0 I}{2\pi h} = \frac{2 \times 10^{-7} I}{h}$$
(13)

The flux density in microtesla is

$$B(\mu T) = \frac{0.2 I}{h}$$
(14)

1000 A is a representative transmission line current, and with h equal to 10 m, the flux density is of order $20\,\mu T$. The simple 'rule of thumb' formula should only be used directly below the conductors, where the lower conductor is much closer to the observer than the upper ones and the field cancellation effect due to these upper conductors is small. When one is at a horizontal distance from the line, the

---

[10] D. L. Henshaw, A. N. Ross, A. P. Fews and A. W. Preece, *Int. J. Radiat. Biol.*, 1996, **69**, 25–38.
[11] J. C. H. Miles and R. A. Algar, *Radiat. Prot. Dosim.*, 1997, **74**, 193–194.

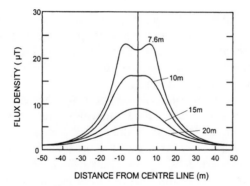

Figure 6 Flux density at 1 m above ground due to a double circuit, 400 kV, line carrying 1000 A on each circuit

contributions of all of the conductors need to be considered. Figure 6 shows the calculated flux densities around a double circuit line carrying 1000 A on each circuit and with a range of conductor heights.

Just as the electric field can induce voltages on objects in the field, an alternating magnetic flux can have a similar effect. Where one has a long run of metal fencing parallel to an overhead line, voltages may be induced which are perceptible to humans and animals. This effect can be eliminated by inserting electrically insulated breaks at intervals and earthing the exposed metalwork.

6 Some Non-field Issues

There is a natural tendency for everything which happens around a line to be blamed on electric or magnetic fields, but two 'non-field' issues deserve a mention. The towers which support overhead wires are large galvanized steel structures. The zinc[12] provides sacrificial protection of the steel and the zinc run-off from the structure can have a detrimental effect on crops and vegetation in the vicinity of the tower bottom. It should be noted that this is the region where fields are at their lowest because the wires are at their highest.

From time to time, birds fly into overhead conductors, both energized and non-energized. Where a line is recognized as being on a flight path, markers are sometimes put on the conductors and/or earth wires to improve their visibility. To avoid corona, only the earth wires are so fitted on transmission lines. Birds do not perch on high voltage conductors because the tips of their feathers would go into corona, but the earth wires which connect the tops of the towers can be popular roosting spots for starlings—with inevitable results for the area below.

7 Studies of Plants and Animals in the Vicinity of Transmission Lines

The New South Wales government[13] held an inquiry into 'Community Needs and High Voltage Transmission Line Development' under the chairmanship of the retired Australian Chief Justice, Sir Harry Gibbs. His report included a review

---

[12] R. Jones, K. A. Prohaska and M. S. E. Burgess, *Water, Air, Soil Pollut.*, 1988, **37**, 355–363.
[13] The Right Honorable Sir Harry Gibbs, *Inquiry in Community Needs and High Voltage Transmission Line Development*, New South Wales Government, Sydney, 1991, ch. 6, p. 69 *et seq.*

of the impact of lines on native flora and fauna, farm animals, and plants. It concluded that 'from a practical point of view, the electric fields created by transmission lines have no adverse effect on crops, pasture grasses, or native flora, other than trees, growing under or near to the lines'. The effect on trees is that due to corona which is described above.

Sir Harry's report drew heavily on the review which was prepared by the Biological Studies Task Team of the Bonneville Power Administration for the US Department of Energy.[14] Many of the team's studies were carried out around 765 kV lines, which have a far higher voltage than those in the UK.

A study over the period 1977–79 of 11 livestock farms crossed by a 765 kV line showed no effect on the health, behaviour, or performance of beef cattle, dairy cattle, sheep, pigs, or horses of the electric and magnetic fields generated by the line. A similar study over six years at 55 dairy farms located near 765 kV lines in Ohio found no effect on health or milk production of the cows.

Pigs were raised in pens housed under a 345 kV lines in Iowa and the experiment showed that the fields from the line had no negative effects on the performance, behaviour, or carcass quality.

The effect of electromagnetic fields on the fertility of cows has been extensively studied in Sweden. The University of Agricultural Sciences in Skara[15] carried out a research project into the fertility of cows which had been exposed to 400 kV lines for long periods. Farms were lined out so that cows were kept in two separate groups during the grazing season, with one group beneath the lines and the other far from them. None of the fertility parameters studied were affected by the exposure, which averaged $4\,\mathrm{kV\,m^{-1}}$ for the electric field and $2\,\mu\mathrm{T}$ for the magnetic one.

The effect of electric fields on tree growth has already been mentioned. The effect is most pronounced in fir trees which have pointed needles, giving rise to large local field enhancements; no comparable effects have been observed on oaks with their rounded leaves. For coniferous trees and plants with tipped leaves, corona effects begin when the unperturbed local field is around $20\,\mathrm{kV\,m^{-1}}$, a far higher value than that to be found at ground level beneath a 400 kV line.

The New South Wales report lists experiments which have been carried out on crops of corn, cotton, soy beans, clover, pasture grasses, and wheat grown near high voltage lines with voltage up to 1200 kV. It concludes that, from a practical point of view, transmission lines have no effect on crops, although they do have an effect on trees which are allowed to grow too close to them. It also concludes that there would seem to be no reason to reach any different conclusion in relation to native flora.

The Bonneville Power report[14] notes that when transmission lines cross open country, birds such as hawks and eagles often use the towers for perching and nesting where they can be exposed to electric fields for long periods. A sample of hawks nesting on 500 and 230 kV lines produced the same average number of young as those nesting in trees and cliffs.

Small mammals were studied for several years as part of the Bonneville Power

[14] Bonneville Power Administration, *Electrical and Biological Effects of Transmission Lines: A Review*, US Department of Energy, Portland, 1982.

[15] Vattenfall, *High Voltage Power Lines, Health and Environment*, Vallingby, Sweden, 1989, p. 30.

research programme on their 1100 kV prototype line and no adverse effects were found. The creatures were most abundant during the first two years of operation when they took advantage of the area of cleared vegetation beneath the line. The vegetation was, however, high enough to shield small mammals from the field.

8 Biological Mechanisms

The data on the biological and health effects of electric and magnetic fields have been reviewed by over eighty national or international scientific bodies. At the time of writing, the most recent such review is that by the International Commission on Non-ionizing Radiation Protection[16] which, as mentioned in the introduction, is the World Health Organization's formally recognized non-governmental organization in non-ionizing radiation. In the summary of biological effects and epidemiological studies (up to 100 kHz) which introduces their recently issued exposure guidelines, they conclude:

- There is currently no convincing evidence for carcinogenic effects of these fields and the data cannot be used as a basis for developing exposure guidelines
- Laboratory studies on cellular and animal systems have found no established adverse effects of low frequency fields when the induced current density is less than $10 \, \mathrm{mA \, m^{-2}}$
- The epidemiological data are insufficient to establish exposure guidelines
- No consistent evidence of adverse reproductive effects has been provided
- Measurement of biological responses in laboratory studies at commonly experienced field levels has provided little indication of adverse health effects of low frequency fields

These conclusions are in line with those of the report on non-ionizing radiation by the European Commission which point out that, whilst there are well-established effects of induced currents on the nervous system, most biological studies suggest that exposure to low frequency fields does not have any significant effect on mammalian development.

A review by a group from the Department of Biochemistry at Cambridge University[17] takes a particularly robust and sceptical view of the biological data; indeed, they write of their recurring theme being the 'overriding need to demonstrate a single, unequivocal ELF-EMF-induced response that will be consistently reproducible in independent laboratories'.

One cannot, of course, say that there are no effects of electric and magnetic field effects, and research continues into possible mechanisms. One such mechanism, which is discussed in the reviews mentioned above, is of particular interest because it was tested (*inter alia*) on sheep penned beneath the conductors of a 500 kV line belonging to the Bonneville Power Administration.[18] Various investigations have shown that power frequency electric and magnetic fields may affect circadian rhythms in animals and a number of studies reported changes in

[16] International Commission on Non-Ionizing Radiation Protection, *Health Phys.*, 1998, **74**, 494–522.

[17] A. Lacy-Hulbert, J. C. Metcalfe and R. Hesketh, *FASEB J.*, 1998, **12**, 395–420.

[18] J. M. Lee, F. Stormshak, J. Thompson, D. L. Hess and D. L. Foster, *Bioelectromagnetics*, 1995, **16**, 119–123.

the function of the pineal gland. It has been suggested that the resulting effect of serum melatonin concentration could contribute to an increased risk of cancer *via* its effects on mammary tumour growth. It should, however, be noted that the ICNIRP publication states that recent studies have found no evidence for a significant effect of exposure to ELF magnetic fields on melatonin levels in humans. In the experiments carried out around the 500 kV line, one group of female Suffolk lambs was penned beneath the line, where the field averaged $6 \, \mathrm{kV \, m^{-1}}$ and $20 \, \mu\mathrm{T}$, and the other was penned 229 m away where the fields were less than $10 \, \mathrm{V \, m^{-1}}$ and $0.03 \, \mu\mathrm{T}$. Melatonin secretion did not vary between the two groups, nor did the age at puberty or the number of subsequent oestrous cycles.

9   Electric and Magnetic Fields and Health

The concerns over the environmental effects of electric and magnetic fields have in large part been driven by the publicity which has been given, quite naturally, to the epidemiological studies of childhood cancers, particularly leukaemia. A number of major reviews of the subject have been published, notably those by the US National Research Council,[19] the UK National Radiological Protection Board,[20] the European Commission,[1] and the International Commission on Non-Ionizing Radiation Protection.[16] The ICNIRP document is the most recent and draws on information from the others.

The hypothesis that magnetic field exposure is associated with childhood cancer mortality was generated by the study carried out in Denver, Colorado, by Wertheimer and Leeper.[21] They found an association between cancer mortality and the 'wire code', which is a measure of a house's proximity to overhead distribution wires; these have a primary voltage of 13 kV in Denver. This association led to the hypothesis that the overhead wire's contribution to the ambient residential magnetic field could be linked to an increased risk of childhood cancer. The findings of this hypothesis-generating study prompted research is several countries, and the American National Research Council review, which was published in 1997, considered 16 such studies from around the world, of which 11 were considered suitable for inclusion in a meta analysis.

This found that wire codes were associated with an approximate 1.5-fold, statistically significant, excess of childhood leukaemia, but that average magnetic field measurements in the home have not been found to be associated with an excess of leukaemia or other cancers. This 'wire code paradox' has led to the development of alternative hypotheses, notably that, because the high wire codes also correlate with traffic density, the agent of interest is traffic fumes rather than magnetic fields.

The International Commission on Non-Ionizing Radiation had a number of additional studies available for consideration in their review which was published

[19] United States National Research Council, *Possible Health Effects of Exposure to Residential Electric and Magnetic Fields*, National Academy Press, Washington, 1997, ch. 5, pp. 117–191.

[20] National Radiological Protection Board, *Board Statement on Restrictions on Human Exposure to Static and Time Varying Electromagnetic Fields and Radiation*, Documents of the NRPB, Chilton, 1993, vol. 4, no. 5.

[21] N. Wertheimer and E. Leeper, *Am. J. Epidemiol.*, 1979, **113**, 487–490.

in April 1998, notably that on childhood leukaemia by the American National Cancer Institute,[22] which, with 638 cases and 620 controls, was far larger than any of the previously published ones.

These data are totally negative as far as wire codes are concerned, but raise some intriguing questions regarding magnetic fields. The authors tested the *a priori* hypothesis that average magnetic fields of more than 0.2 μT were associated with leukaemia and concluded that they were not. However, if higher 'cut points' are used, the data, in the opinion of the ICNIRP review, are suggestive of a positive association. In total, though, the review concludes that, in the absence of support from laboratory studies, the epidemiological data are insufficient to allow an exposure guideline to be established.

10   Limitations on Exposure

In the UK, the responsibility for advising on limitations on exposure to electric and magnetic fields rests with the National Radiological Protection Board.[20] The Board established an Advisory Group on Non-Ionizing Radiation which reviewed the evidence for an association between electric and magnetic field exposure and cancer.

The review body concluded that there is no clear evidence of adverse health effects at the levels of electric and magnetic fields to which people are normally exposed. Like ICNIRP, they concluded that the epidemiological data do not provide a basis for restricting exposure. The Board's guidance is therefore based on the available biological data describing thresholds for well-established direct and indirect effects of acute exposure.

For 50 Hz fields, the Board sets a 'basic restriction' of $10\,\mathrm{mA\,m^{-2}}$ for the current induced in the head and trunk by the field. It then defines an 'investigation level' for the purpose of comparison with measured fields. If the measurement is less than the investigation level, compliance with the basic restriction is assured. Should the measured value be greater than the investigation level, it does not follow that the restriction is violated. The high field, for example, may be highly localized around an electrical appliance and the volume of the high field area may not be sufficient to generate currents which exceed the reference in the body. As its name implies, if the investigation level is exceeded, one needs to investigate whether the basic restriction is exceeded as well. The investigation level is not a limit.

For 50 Hz, the investigation levels are:

- Electric field $12\,\mathrm{kV\,m^{-1}}$
- Magnetic field $1600\,\mu$T

11   Conclusions

Overhead lines have an obvious visual impact on the environment, but their electric and magnetic fields have a limited ecological effect. If trees are allowed to

[22] M. S. Linet, E. E. Hatch, R. A. Kleinerman, L. L. Robison, W. T. Kaune, D. R. Friedman, R. K. Severson, C. M. Haines, C. T. Hartstock, S. N. Niwa, S. M. Wacholder and R. E Tarone, *New Engl. J. Med.*, 1997, **337**, 1–7.

grow beneath overhead line conductors, then unrestrained growth will be limited by the onset of corona. However, for reasons of safety, the trees need to be cut back before this becomes a problem.

If bee hives are sited in the electric fields below high voltage lines, then output may be affected because the bees experience repeated small shocks as a consequence of electrostatic induction. This effect can be eliminated by screening the hive with wire mesh. Animals may also experience shocks due to induced voltages on drinking troughs and fences, but these effects can be minimized by earthing the offending item.

Studies carried out in the USA and Sweden indicate that farm animals and crops are not adversely affected by the presence of high voltage overhead lines.

Energy Efficiency and Conservation

ANDREW WARREN

1   Introduction to Least Cost Planning

Introduction

For energy suppliers, planning is all about making rational economic choices. Classic economic theory states that energy producers will plan and invest in the method of production that maximizes their profits, while consumers will spend such that they maximize the benefits of their expenditure. In a competitive market, supply and demand would operate such that goods and services will be produced at the least possible cost and sold at the lowest price. In such an ideal world the investment plans of an energy supplier will be *ipso facto* 'least cost' plans from a societal perspective.

However, economic theory and economic reality are not, unfortunately, one and the same. Energy markets are imperfect markets, causing investments in energy supply and energy use to be made on the basis of completely different criteria. This analysis of available options has to date been exclusively concerned with the supply of energy. 'Demand' is the sum of millions of end uses: heating, appliances, motors, and lighting. In all of these end uses, energy efficiency can be improved through programmes which can be run by energy suppliers.

Thus least cost planning involves viewing demand not as a given, but as a variable, since the efficiency with which energy is used in the multiplicity of end uses can be improved, through utility investment in energy management programmes. In just the same way that supply investments deliver a certain amount of gas or electricity for a certain cost, energy management programmes can deliver and use efficiency improvements—'saved' energy—for a certain cost.

The market imperfections and barriers which hinder optimal investment in energy efficiency include:

- *Lack of knowledge.* Many energy customers are not aware of the potential savings that can be achieved from energy efficiency, nor do they have the knowledge and technical skills to install energy efficiency measures.

Issues in Environmental Science and Technology, No. 11
Environmental Impact of Power Generation

- *Limited access to capital.* Energy efficiency measures often require up-front capital in order to achieve long-term savings. Domestic customers and small businesses often do not have the up-front capital, or they reserve it for other investments.
- *Rapid payback requirements.* Extensive surveys have indicated that domestic, commercial, and industrial customers will not undertake investments in energy efficiency unless the pay-back period is as short as six months to three years. Electric utility investments in generation plants, on the other hand, are made with expected pay-back periods of 10 years or more. This creates a significant imbalance in favour of supply-side investments.
- *Lack of responsibility.* Electricity consumers that rent their homes or offices do not have an interest in long-term investments necessary to improve the energy efficiency of buildings, because they do not own the buildings. On the other hand, landlords do not have an interest in such investments because they do not pay the electricity bills.
- *Inappropriate price signals.* Electricity rates often do not reflect the full marginal cost of producing electricity (especially the environmental costs), nor do they fully reflect the variation in costs between peak, shoulder, and off-peak periods.
- *Lack of rational decision-making.* The electricity bill is often a small part of a customer's overall budget, and rarely receives appropriate attention. In addition, customers are often governed by unrelated, non-economic issues, such as appearance, fashion, trends, and momentum of habit.
- *Lack of access to and trust in efficiency equipment.* Because some energy efficiency measures are relatively new to the market, they can be difficult to purchase, and they can be viewed as too risky by many customers.

What is Least Cost Planning?

The concept of least cost planning was originally mooted by a former director of the Office of Conservation at the US Department of Energy, Roger Sant. Sant's definition of least cost planning, a definition which has been accepted by many utility regulators in the USA, is as follows:

"The 'least cost strategy' . . . provides for meeting the need of energy services with the least costly mix of energy supplies and energy efficiency improvements."

The starting point of Sant's analysis is the concept of energy *services*. Customers do not require energy as a product *per se*, but rather they require electricity and gas for the services which they can provide. Consumers do not need kilowatt hours or therms but the heat, light, and mechanical drive, *i.e.* the 'energy services' which the energy produces.

Increases in the demand for energy services can be met by increased energy supply, by improved energy efficiency of energy-using equipment, or by a combination of both. Sant's definition of the least cost energy strategy is the mix of energy inputs (supplies) and energy efficiency improvements having the lowest total cost to society. Least cost planning, therefore, is the planning process

Figure 1 A simplified schematic of the least cost planning process

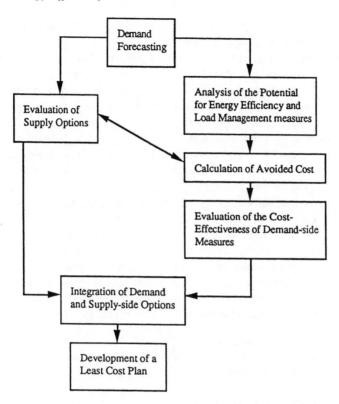

necessary to achieve this lowest cost. It should be noted, however, that Sant's definition is based on the societal perspective, rather than on the more narrow perspective of any specific group.

However, many utilities in the USA, while supporting this broad aim, are unhappy with the term 'least cost planning' since the term implies that there is *one* optimal solution which will be *the* least cost solution. For this reason, many utilities are more comfortable with the term 'integrated resource planning' (IRP). The European Commission has further evolved the terminology, encapsulating the concept as 'rational planning techniques', the title given to its 1996 draft directive. The differences in terminology is less important than the observation that the common element is the integration of both supply side and demand side by comparing the costs of these options.

It is beyond the scope of this article to give a detailed description of the methodology of least cost planning, but the diagram in Figure 1 sets out, in a simplified form, the way in which least cost planning is achieved in practice.

IRP can be divided into two different, but overlapping, aspects. The first is the *practices* used by electric utilities in planning for future energy resources. These practices are used by utilities to achieve the general objectives of IRP. They typically involve detailed planning methods such as demand and supply forecasts, energy technology assessments, economic analyses, and utility modelling techniques.

The second aspect of IRP is the *policies* used by government agencies to

115

promote appropriate practices by utilities. Such regulatory policies are necessary to encourage, support, and guide utility IRP practices, and in the USA vary widely among state regulatory agencies. However, there are certain common elements in each state that are critical in promoting IRP.

It is important to note, however, that IRP should also be applied to the gas industry. Ideally, resource planning procedures should integrate gas, electricity, and other energy opportunities, in order to optimize total energy use. This is especially important in Europe because of a growing trend for distribution companies to become horizontally integrated (*i.e.* supplying electricity, gas, and heat) as a means of improving both operational efficiency and energy efficiency. In general, the arguments, policies, and recommendations made in this article regarding the electricity industry apply to natural gas utilities as well.

Improving energy efficiency is a central component of IRP. Many European countries currently have a number of government programmes and policies to encourage energy efficiency. These range from energy efficiency standards, to time-of-use tariffs, to customer subsidies for energy efficiency investments.

Within this concept, utilities change their business planning to provide least cost energy services to their customers; the focus is on low bills for energy services, not on low prices for the unit (kW h) of energy. The utilities can achieve this with

- Targeted consulting
- Rebates for energy efficient appliances and equipment
- Free installation of energy saving items
- Third-party financing
- Demand-side bidding

Each of these options is explored in greater depth later in this article.

Thus the term 'planning' does not imply that state authorities are to impose plans onto utilities. Under the draft directive, EU member states, instead, have to provide a framework which makes it legally and economically viable for distribution/supply utilities to include all types of resources which have environmental and net economic benefits to society into their business plan. Under traditional price setting or regulation, there is an incentive for utilities to sell more, not less, kW h, even if reducing demand through energy efficiency costs much less than generating and supplying additional energy.

There are, in principle, two ways to alleviate or remove this disincentive for utilities to start broader energy end-use efficiency activities:

- Utilities offer energy services (*e.g.* efficient lighting) instead of energy (kW h) and receive the revenue directly from the customer through concepts like third-party financing or leasing schemes
- Utilities are allowed to recover costs for demand-side energy efficiency programmes through energy prices and tariffs as long as customers' average bills decrease. These programmes do not have to be performed entirely by the utility itself. On the contrary, by co-operating with market partners like manufacturers, trade, craftsmen, and engineering consultants, utility energy efficiency programmes as well as energy services open up the market for providers of energy-efficient technologies

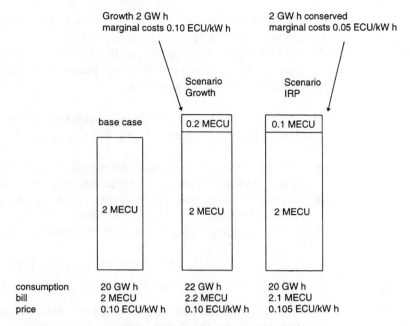

Both possibilities are addressed by the proposed Council directive on Rational Planning Techniques (RPT) (Article 2c, 2d), which was approved by the European Parliament with some amendments but with an overwhelming majority in November 1996. Most amendments have been included in the amended proposal prepared by the European Commission in March 1997.

Clearly it is justifiable to finance energy efficiency programmes through marginally increased rates, just like power plants and transmission/distribution lines, so long as the total costs to customers are reduced through the 'conservation power plant'. Owing to lower kW h sales, reduced costs will have to be spread over fewer kW h. This may lead to a slight increase in prices per kW h, but even then, total costs to customers for the energy services they require will still decrease since the number of kW h consumed has decreased.

Figure 2 provides a very simple schematic example of these effects on bills and prices: starting from the base case (left column), it is assumed that during a certain period an increase in demand of 10% is expected. This demand can either be met by increasing generation ('Scenario Growth'), or the additional demand can be avoided by energy efficiency programmes ('Scenario IRP'). When both scenarios are compared it becomes clear that the total costs of energy services are lower in the IRP scenario, but that unit prices have to be higher in this scenario since the number of kW h sold is lower too.

Therefore, the measure of economic efficiency changes with the introduction of RPT. It is no longer the price per kW h, but the total costs for energy services, which is effectively of far greater concern to consumers. Thus, IRP/RPT is an instrument where the economic sectors causing the pollution are the actors for avoiding it. RPT can be regarded as an energy pricing policy instrument in two senses:

(i) RPT/IRP is a way to directly internalize the costs for *avoiding* energy-related emissions into the prices of the same energy

(ii) RPT/IRP reduces energy *bills* and thus increases industry's competitiveness and private customers' purchasing power, but energy *prices* rise very modestly (a few percentage points)

In order to make energy efficiency programmes economically viable and attractive for utilities, incentives are needed:

- Approval of the costs of demand-side management (DSM) activities as costs of a rational and efficient management of electric distribution/supply utilities
- Mechanisms to neutralize the incentive to exceed the sales forecast and to recover those fixed costs in the case of falling short of the sales forecast which the utility cannot avoid in the short and medium run ('Decoupling Sales and Profits')
- Positive incentives that are able to motivate the distributors/suppliers for IRP/DSM activities

While the first two elements are intended to eliminate the most important negative incentives for IRP/DSM activities, the third element gives an additional positive signal.

However, where the functions of distribution and supply are separated, the suppliers will not be motivated to increase rates, even for very cost-effective energy efficiency programmes, because they will be afraid to lose customers to 'cheaper' suppliers. Distribution utilities, on the other side, could receive regulatory incentives but have no direct contact with customers.

One possibility to harmonize both monopoly and competitive situations would be to introduce levies of 3–5% on the price of electricity and gas, leaving utilities the choice of whether they want to raise the levies, or to undertake their own NEGA Watt programmes in which they invest all or part of the amount of levies they would have to raise.

2 Integrated Resource Planning (IRP) as an Instrument of Strategic Utility Planning

As shown in the previous section, the existence of cost-effective energy efficiency potentials is the reason and the starting point for making IRP an instrument of strategic utility planning. Whenever end-use energy efficiency is cheaper than production and supply of end-use energy, it should be made use of as a resource ('NEGA Watt'). Utilities should encourage and support consumers to use energy-efficient end-use technologies in order to achieve lower overall costs of energy services. By doing so, utilities should try to achieve quantitative targets of conserved energy and capacity, which are determined through a cost–benefit analysis during an IRP process. This means going beyond the generalized information and advice that utilities have long given to their customers.

There are five ways to deliver such targeted support in DSM activities and programmes targeting specific customer sectors and end-use technologies:

(i) *Consulting.* By giving energy audits at the customer's premises free of charge or at a low fee, utilities can help customers detect cost-effective energy efficiency potentials. This can also be done by consultants which are paid for by the utilities. Danish utilities have conducted or paid more than 4000 audits in industrial and commercial companies and in public institutions during recent years. On average, about 10% savings potential with an average pay-back of two years have been detected, of which 45% have been implemented by customers.

(ii) The rate of implementation of energy efficiency measures detected during an audit may be increased if a *rebate* on the investment is offered together with the audit. This may be a fixed amount for standard devices like high-frequency ballasts. Such fixed rebates are also very common among German utilities for electric household appliances. RWE Energie, the largest German utility, paid one million rebates of 100 DM each for efficient refrigerators, freezers, dish-washers, and washing machines between 1992 and 1995. The programme saved 450 GW h of electricity; the costs to society were below the long-run marginal costs of electricity supply. For larger investments in individual technical solutions, a rebate based on the predicted yearly savings is appropriate, *e.g.* the 'customized rebate' programme by the largest Californian utility, Pacific Gas & Electric.

(iii) With *direct installation* or free give-away programmes, utilities offer comparatively cheap devices like compact fluorescent lamps (CFLs) or low-flow shower heads to their customers for free. These programmes can be very effective, reaching high participation rates, and also often rather cheap. Many European utilities have offered CFLs free to their residential, and sometimes small commercial, customers. One example is the programme of Stadtwerke Saarbrucken, a municipal utility serving 100 000 customers, implemented in 1994. More than 50% of the residential customers came to the advice centre of the utility to get their free lamp. About 2 GW h were saved cost-effectively to the utility and the society. This success was possible because the offer of the free lamp was combined with an advertizing campaign making people familiar with CFLs. Utilities in the USA are also offering such free services to small commercial customers.

(iv) An even bigger financial engagement is required with *third-party financing*. On the other hand, third-party financing allows the utility to recover its expenditures directly from the individual customer *via* bilateral contracts, while expenditure for other types of programmes with audits and/or financial incentives have to be recovered through rate from all customers. Many European utilities are examining ways to expand their business to the demand side and introduce energy-efficient services through third-party financing, and there are now a lot of pilot schemes under way.

(v) An even more innovative and market-oriented possibility to include also large customers and private energy service companies into the efforts to realize cost-effective electricity savings is *demand-side bidding.* In DSM bidding schemes, the utility publishes the need for saved capacity and electricity and asks for offers. Large industrial customers or consultants who have the technical skills to develop projects for electricity efficiency can offer these projects to the utility and ask for a certain payment per certified kW of kW h from the utility. The utility chooses from the projects up to a certain limit of capacity or a certain price cap and makes contracts with the successful bidders to provide and proof the savings. DSM bidding has become increasingly common in the USA during recent years but has not yet been tried in Europe.

Not every activity connected with DSM programmes of these types has to be carried out by the utility itself. The utility will probably have neither the capability nor the capacity to perform a large number of energy audits on all technical details of efficient electric end-use technology. The utility, however, should provide the financial and organizational framework, thus creating an opportunity and a market for engineering consultants, technical contractors, and customers to carry out more energy efficiency measures than they would do without DSM programmes.

Implementation of IRP in the USA has resulted in a substantial increase in the efficiency with which energy is used, has saved billions of dollars for electric utilities and consumers, and has reduced the use of environmentally threatening power plants. For example:

- Southern California Edison estimates that its energy efficiency programmes will reduce its total electricity demand in the year 2000 by 13%
- The Pacific Gas and Electric Company announced in its annual report that it plans to meet 75% of all new demand this decade with energy efficiency programmes, and that the remaining 25% will be met with small-scale independent power projects, including renewable resources
- The Northwest Power Planning Council plans to develop 1500 MW h of energy efficiency savings by the year 2000 for about half of the cost of providing the power from conventional resources
- A recent census found that USA electric utilities spent $1.2 billion on energy efficiency programmes in 1990, with estimated savings of 24 400 MW h
- One analysis forecasts that if current IRP trends continue, energy efficiency programmes could reduce total USA electricity consumption by 20% by the year 2010, cutting electricity bills by $61 billion per year, and reducing total USA CO_2 emission in that year by 9%.

Costs and Benefits of IRP/DSM

In general, the *costs* of an energy efficiency programme include the costs to the utility to market, deliver, and install energy efficiency measures. The *benefits* include the utility's 'avoided costs', which are the costs of the generation resources that are avoided as a result of energy and capacity saved by the programme. Avoided costs include *short-term costs*, such as the capital costs of generation facilities or the fixed component of purchases.

A programme is considered cost-effective if its benefits outweigh its costs. This

Table 1 Different perspectives for assessing DSM cost-effectiveness

	DSM benefits				DSM costs		
	Utility avoided costs	Customer bill savings	Utility incentive payment	Avoided externalities	Utility programme costs	Customer costs	Customer bill savings
Participating customer		X	X			X	
Non-participating customer	X				X		X
Utility	X				X		
Society	X			X	X	X	

is usually expressed if programmes have a positive net present value, or a benefit-cost ratio in excess of one. All costs and benefits should be estimated over the total lifetime of the energy efficiency measures, and should be compared in terms of present values.

Costs and Benefits from Different Perspectives

DSM programmes differ from generation facilities in that certain customers often pay for a portion of the costs of installing energy efficiency measures, and these customers also receive additional benefits in the form of lower electricity bills. In other words, the costs and benefits of the programmes are shared differently between the utility and the customers. As a result, the cost-effectiveness of an energy efficiency programme will vary, depending upon whose perspective is considered in the analysis. DSM cost–benefit analyses can be conducted to address four different, but overlapping, perspectives: the customer who participates in the DSM programme; the customer who does not participate; the utility; and society. The different costs and benefits of each perspective are summarized in Table 1, and are discussed briefly below.

Participating customer. The goal of applying this perspective is to determine the impact on the customer who participates in the DSM programme. The participant's costs will include all costs that he/she incurs in financing, installing, and operating a particular efficiency measure. The benefits will include the savings on the participant's electricity bills, as well as any incentive paid by the utility to assist the participant in the financing, installation, or operation of the measure.

Non-participating customer. The goal of applying this perspective is to identify any distribution or equity impacts from the DSM programme, *i.e.* to determine the extent to which non-participating customers pay for, or benefit from, the programme. The non-participant will receive benefits from the DSM programme in the form of the utility's avoided costs, *i.e.* the reduction in costs to generate power. The costs to the non-participant will include the utility's programme administration costs plus the participating customer's bill savings, because these impacts will eventually be reflected in the electricity rates. Evaluating programmes from this perspective is referred to as the 'no-losers' test, because it limits programmes to those that will not raise rates.

Utility. The goal of applying this perspective is to identify the impact of the DSM programme on a utility's system costs. Evaluating programmes from this perspective is consistent with conventional practice for evaluating supply-side resources. The costs include all utility programme administration costs; these include utility staff costs, marketing and advertising costs, energy auditing costs, financing costs, or any incentive payments necessary to induce customers to participate in the programme. The benefits include the utility's avoided costs, *i.e.* the reduction in costs to generate power.

Society. The goal of applying this perspective is to identify the impact of a DSM programme on society, regardless of the precise allocation of costs and benefits to the utility, the ratepayers, or the programme participants. In addition, this perspective recognizes that there are additional costs and benefits (*e.g.* environmental benefits) of DSM programmes which are not addressed when the perspective is limited to the utility and its customers. Therefore, under this perspective, DSM costs include the utility's programme administration costs, plus the customer's cost to install and operate any efficiency measure. The DSM programme benefits include the utility's avoided costs, plus any additional benefits to society such as avoided environmental impacts.

It is to the assessment of the best means of measurement of such environmental impacts that the rest of this article is devoted.

3 Accounting for the Environmental Externalities of Electricity Production—A Summary of USA Practice

The production and consumption of electricity imposes a number of costs on society. Some of these costs are paid by the electricity utilities and their customers, while others are not accounted for in the electricity marketplace. These latter costs are referred to as 'externalities', because they are external to market transactions and prices. Environmental impacts are the most prominent example of the externalities of electricity generation.

In recent years, regulatory agencies in the USA have increased their oversight of the long-term energy planning practices of electricity utilities. This has generally resulted in consideration of a broader range of energy options, such as energy efficiency and other less-conventional resources, as well as consideration of all of the various costs and benefits of those resources. As a result, a number of state regulatory agencies have established policies which require utilities to account for environmental externalities when selecting their future resource options.

These policies have been implemented in a variety of forms. Some of the more cautious policies simply require utilities to make qualitative, judgmental decisions about environmental impacts. Others more specifically require utilities to estimate environmental costs in monetary terms, and to include those costs with the conventional costs when selecting among resource options. The overall goal of these policies is to encourage utilities to select electricity resources which result in the lowest overall costs to society.

Estimating monetary values of environmental externalities is often seen as the preferable method for determining the total costs to society, because it places these costs in a unit of measurement that is common with the conventional costs of generating electricity. In other words, it allows for a direct comparison of economic and environmental costs. However, estimating monetary values of environmental externalities is an extremely difficult task because of the many uncertainties involved, and because of the different values placed on the environment by different parties. Therefore, much of the debate in the USA has focused on the practice of estimating monetary values of environmental costs, as well as the policies necessary to account for them.

The purpose of this part of the article is to summarize the practices and policies being used in the USA to account for the environmental externalities of electricity generation. Some of the results and lessons from this experience should prove useful in developing environmental policies in the European Community.

Methods of Estimating Monetary Values of Environmental Externalities

USA regulatory agencies use a variety of techniques to account for environmental externalities of electricity production. These range from qualitative, subjective techniques to those which rely upon monetary values of environmental impacts. The use of monetary values is becoming more widely used, and is seen as being the most effective technique for addressing environmental impacts. However, this approach is also the most controversial and complex. Therefore, the primary methods of estimating monetary values of environmental externalities are described here as background.

The two primary methods of estimating monetary values of environmental externalities are the 'damage cost' approach and the 'control cost' approach. Damage costs are developed by identifying the amount of environmental damage, in terms of land used, wildlife effects, or human health impacts, and then placing a monetary value on that damage. Alternatively, control costs are developed using the costs to reduce pollution emissions or to mitigate environmental damage.

4   Monetization Using Damage Costs

Identification of Environmental Damage

The damage cost approach to monetizing environmental impacts requires two general tasks: identifying the environmental damage caused by electricity production, and then estimating the value of that damage. The first task can be further broken down into four steps:

- Determining the scale of *emissions* of a pollutant (*e.g.* pounds of SO_2)
- Determining the *dispersal* of the pollutant (*e.g.* where does the SO_2 go, and how does it react with other elements to form acid precipitation)
- Estimating the size and type of the *population exposed* to the pollutant (*e.g.* which people, trees, lakes, or crops does the acid precipitation land on)
- Estimating the *response* of each population exposed (*e.g.* how does acid precipitation affect human health, flora, fauna, buildings)

Identifying environmental damage through these four steps is an exceptionally complex task, loaded with uncertainties. For the first step, the scale of emissions from plant operation is generally well known for most sources. However, pollutant emissions from other components of the fuel chain (*e.g.* fuel extraction, waste disposal) are much less well understood.

The last three steps depend on myriad factors relating to plant characteristics, atmospheric, biological, chemical, geophysical, ecological, and physiological conditions and relationships. Each step requires a great deal of data for modelling, and contains many uncertainties that may not be predictable or well understood. In addition, the conditions are likely to vary significantly over time (throughout a day and throughout years), and will depend upon location as well as socio-economic and cultural conditions.

Valuation of Environmental Damage

Determining the monetary value of environmental damage can be even more challenging and controversial than estimating the damage itself. The very nature of human health and the environment is such that determining values requires subjective judgements that will vary by many factors. It can be argued that many resources, such as sacred religious sites or endangered species, are simply not priceable. Nevertheless, economists have developed a number of methods of monetizing non-monetary goods. A brief description of these methods and their limitations is given below.

Market-based Values

For environmental goods that are traded in a market (*e.g.* fish, timber, agriculture), it is possible to estimate the monetary value of the damage of those goods by multiplying the quantity of goods lost by the market prices. However, market prices are not necessarily a good indication of the value that society places on these environmental elements, for a number of reasons. First, market imperfections can distort prices. Second, prices do not reflect values people might place on the 'option value' of a resource (*i.e.* the option to use an environmental resource at some point in the future) or on the 'existence value' of a resource (a desire to preserve a resource, even if an individual has no plans to use it). Third, prices do not reflect the unique quality of resources or the irreversibility of certain losses.

For non-market goods there are a variety of methods that can be used to monetize environmental resources. The two most common are revealed preference methods and contingent valuation techniques.

Revealed Preferences

Revealed preference methods infer what value people place on goods and services by observing their behaviour. A common example is to estimate the costs people are willing to incur to travel to a certain facility (*e.g.* a recreation site). The primary costs may include transportation costs and the cost of time spent, represented by the wage of the individual. The travel cost method is subject to many fundamental difficulties, including: lack of data, variations across households of wage rates and preferences, variable recreation site characteristics, the exclusion of option value or existence value, and others.

Hedonic pricing techniques are another type of revealed preference method. These techniques look at the difference in prices of market-based goods to determine how much people will pay to obtain (or avoid) certain environmental or health benefits (or costs) associated with those goods. For example, wages of workers who face occupational risks can be compared to those who do not face such risks. The difference can be seen to represent the value of that occupational risk, assuming that all other factors are equal or are accounted for. As another example, changes in property values due to nice views or nearby health threats can be used to determine the value of the view or the health threat, again assuming that other factors are accounted for.

Hedonic pricing techniques suffer from a number of systematic limitations and uncertainties. First, it is often difficult to separate the risk or cost in question from the many other factors that are considered in people's decisions to accept jobs or buy homes. Second, people do not have complete freedom to choose any job or home; they are often subject to financial, social, and locational constraints. Third, people often do not have complete information about the potential costs or risks that they face at home or on the job. Fourth, there are data procurement and measurement problems for prices, wages, and environmental and health risks. Fifth, multiple regression techniques are often used to assess relationships, and these techniques are subject to limitations in defining the form of the model, specifying the appropriate variables, and minimizing correlation between the variables.

Contingent Valuation

Contingent valuation techniques are often used to place a monetary value on human life, health, and environmental resources for which there are no market indications of their value. These techniques involve surveying a sample of the population to estimate how much people are willing to pay (WTP) to avoid an undesirable effect, or how much they are willing to be compensated (WTC) if an undesirable outcome occurs. This technique has proved useful because of its simplicity and ability to address many environmental concerns with a consistent methodology.

However, contingent valuation suffers from a number of systematic problems. It is questionable whether people's responses truly reflect their values. They may adjust their responses depending upon whether they actually expect to be paid or to be compensated. One of the first studies to estimate both WTP and WTC found that people's willingness to be compensated was about four times greater than their willingness to pay. There is also a great risk of bias, depending upon how the survey is designed and conducted. This technique also depends upon the respondents being fully informed about the environmental and health risks of the issues in question. Contingent valuation results will vary across people of different incomes, because of the different values that are placed on money. In addition, it is difficult to apply contingent valuation methods to large-scale environmental issues because of the vastly different impacts that will be felt by different people. For example, it is likely that wealthy people in Amsterdam are willing to pay more to avoid global warming than are low-income people of mid-western Canada. Aggregating such preferences would dilute the results.

5 Monetization Using Control Costs

There are a number of instances where regulators require that industry takes steps to limit pollution emissions to an acceptable standard. Such requirements are based on an implicit assumption that the value of protecting the environment equals or exceeds the costs required to protect the environment. Therefore, the costs of controlling emissions can be seen as representing the monetary value that society places on the environment. For example, acid rain regulations that require flue gas desulfurization (FGD) to be installed on coal plants to limit SO_2 emissions can be used to determine a monetary value for SO_2 emissions. The cost of the FGD can be divided by the amount of SO_2 removed to establish a value of \$/tonne of SO_2. This value represents the cost of SO_2 emissions from plants without FGD, as well as the *residual* SO_2 emissions from plants with FGD installed.

It is important to distinguish damage costs from control costs. Damage costs represent the costs to society of particular environmental impacts. Accordingly, damage costs represent the *benefit* to society of avoiding the environmental impact, *i.e.* they represent the monetary benefit of environmental protection. Control costs, on the other hand, represent how much society has to pay to avoid the environmental impact, *i.e.* the monetary *cost* of environmental protection.

When control costs are used to represent environmental externalities, there is an explicit assumption that regulators have set environmental standards to that the costs of the regulations roughly equal the benefits. In other words, it is assumed that regulators set environmental standards at the point where damage costs roughly equal control costs. This assumes that the regulators are well informed and free of political and other constraints in setting environmental standards. Using control costs to estimate environmental externalities is sometimes described as using the 'revealed preferences' of regulators.

It is important to note that only the *marginal* cost of control provides an indication of what society is willing to pay for environmental protection. The marginal cost represents the highest price that society is willing to pay to avoid environmental damage through regulation or emission targets, and therefore is the best representation of the value that society places on that environmental damage. It is irrelevant that lower cost methods (*e.g.* energy efficiency) may exist to avoid some of the environmental damage.

Using control costs to monetize environmental externalities has a number of conceptual and practical difficulties. The most obvious is the assumption that regulators are fully aware of the economic costs and environmental benefits of the regulation that they design. It also assumes that regulation is designed without political or social constraints, including, for example, pressure from the very industries that the regulations are targeted upon. In addition, society's values can change over time, whereas regulation can take years to establish. The control cost approach is also limited by the difficulty of defining the marginal unit of pollutant control. This task is made complex because regulatory requirements can be inconsistent, the marginal unit of pollutant is often in flux, and some externalities may have multiple environmental effects and be subject to multiple regulations.

In spite of the limitations of the control cost technique, it is becoming more accepted as a reasonable 'starting point' for rough estimates of environmental costs, until better damage cost methodologies can be developed. Electric utility regulators in California, Massachusetts, Nevada, New Jersey, and New York have adopted monetary values for environmental costs based on the control cost technique, on the basis that direct monetization of damage costs is not feasible at this time.

Using Control Costs to Monetize CO_2 Emissions

It is now widely accepted that greenhouse gases may impose substantial environmental costs on society. However, because this has only been recently accepted, there are few regulatory standards to limit greenhouse gas emissions. In addition, there is currently no economically practical control technology that can be used to reduce CO_2 emissions from a power plant in the same way that, for example, flue gas desulfurization can be used to reduce SO_2 from a coal plant. As a result, using control costs to monetize CO_2 control policy will need to be established in the near future.

In addition, although there is no end-of-pipe control technology for CO_2 emissions, there are many options available to stabilize the CO_2 emissions (*e.g.* energy efficiency, fuel switching), as well as policies to increase CO_2 sinks (*e.g.* halting the destruction of rainforests, planting additional trees). Therefore, these options can be seen as CO_2 control options, and their costs represent CO_2 control costs.

Under ideal circumstances, a control option 'supply curve' could be constructed to identify the marginal CO_2 control option. The control option supply curve would present the costs of all the CO_2 control options, along with the potential quantity of CO_2 that could be abated (or absorbed) by each option. The control options could then be ordered on the basis of costs. The marginal control option would then be that option that would be sufficient to meet CO_2 emission limits or targets, assuming that all the lower cost options are implemented as well.

Unfortunately, the information to perform such an analysis is not easily available. The costs of the various control options have not been thoroughly estimated, and comprehensive emission limits and targets have not been established yet, especially in the USA. However, analysts have bypassed these limitations by assuming that many measures will be required to address global warming, that tree planting to absorb carbon is likely to be among these measures, and that the cost of planting is likely to represent the marginal cost of any aggressive CO_2 mitigation strategy. Therefore, marginal tree-planting costs have been recommended by many analysts to represent the monetary value of CO_2 emissions.

Some CO_2 cost estimates have been based on the costs of tree-planting programmes in USA urban regions, in USA forests, and in less developed countries such as Costa Rica and Guatemala. Tree-planting programmes located in the USA or in less developed countries could be funded and implemented by any country; therefore, they can represent CO_2 costs for any country, regardless of its location or geographical makeup.

6 Damage Costs *versus* Control Costs for Policy Purposes

When control costs and damage costs are used to represent social cost in evaluating environmental policies, it is important to distinguish between the two. As described above, damage costs represent the benefits of environmental protection, while control costs represent the costs of environmental protection. For many pollutants, the costs and benefits of environmental protection are assumed to be the same, based on the premise that well-informed, unconstrained regulators establish environmental levels to meet this goal.

However, because of the uncertainties about global warming impacts and the lack of comprehensive policies to address them, control cost estimates for CO_2 do not necessarily need to be based on the assumption that control costs equal damage costs. Instead, it is assumed that there is a marginal control option which can be used to avoid global warming problems, regardless of what the full damage costs of CO_2 might be. It is quite possible that there is a divergence between marginal control costs and actual damage costs of CO_2 emissions. This could occur because of the enormous magnitude of the damage that may result from global warming, and because of the lack of political will to establish the necessary standards and practices to prevent it.

For example, consider a CO_2 control cost estimate for a European country based on the European Commission's proposed carbon tax. The carbon tax could be chosen because it represents the European Union's 'revealed preference' for addressing global warming. However, if this control policy does not result in enough CO_2 reduction to mitigate sufficiently the greenhouse effect, then the control cost estimate will be lower than the actual damage cost, and this will result in an underestimate of the social value of CO_2 costs. Therefore, when monetary values are applied to environmental externalities it is important to recognize when there may be a divergence between damage costs and control costs.

This divergence is important because it may result in an under- or over-estimation of the social value of the environmental impact. Furthermore, environmental policies will be most cost-effective (*i.e.* economically efficient) when the costs of environmental protection are set equal to its benefits. Therefore, environmental policies will be most cost-effective to the extent that policy makers understand *both* the damage costs and the control costs of environmental protection. Therefore, efforts should be made to develop a better understanding of both control costs and damage costs. Rather than *assume* that environmental policies are established by well-informed, unconstrained policy makers, it would be preferable to take steps to *ensure* that this occurs.

7 Unpriceable Environmental Costs

There are some environmental costs and risks which will defy economic evaluation, even by the most ardent of economists. A common example of an unpriceable cost is the potential loss of a site of religious, cultural, or historical significance. Policy makers may decide that it is inappropriate or impossible for society to place a value or a price on such costs.

Some analysts argue that unpriceable costs should be accounted for outside of the quantitative economic analysis. In other words, these costs can be treated as planning constraints, within which economic decisions must be made. In the case of electricity resource decision-making, this is consistent with other planning constraints that are currently used, such as ensuring a reliable supply of electricity. The decision-making process will be simplified if most environmental costs can be addressed through the application of monetary values, and then the only remaining planning constraints will be for unpriceable costs. It is important that policy makers identify such planning constraints, and ensure that sufficient opportunities exist for such constraints to be considered in the decision-making process.

Although unpriceable costs may be considered beyond economic evaluation, it must be recognized that when an economic decision is based on such costs, there will be an *implicit* value placed on them. For example, if a $500 million power plant is constructed instead of an otherwise equivalent $450 million plant in order to avoid damage to a religious site, then the decision places an implicit value of $50 million on that site. Conversely, if the $450 million plant were constructed because of its lower costs, then the decision places an implicit value less than $50 million on the site.

8 References

S. Brick and G. Edgar, *Blunting Risk with Caution: the Next Step for Least Cost Planning*, 1990.
S. Buchanan, *Estimating Environmental Costs of Energy Resources*, 1990.
P. Chernick and E. Cavehill, *The Value of Environmental Externalities in Energy Conservation Planning*, 1990.
Elsam (eds.), *Integrated Resources Planning: From Concept to Practice, IRP in the Danish Electric Utilities*, 1994.
European Commission (COM) (97) 69 final
European Parliament Energy Pricing Policy: Targets, Possibilities and Impacts, 1998.
C. Goldman and M. F. Kito, *Review of Demand Side Bidding Programmes*, 1994.
Greenpeace International, *Integrated Resource Planning—Making Electricity Efficiency Work in Europe*, 1993.
E. Mills, *Evaluation of European Lighting Programmes*, 1991.
D. Pearce, *et al.*, *Blueprint for a Green Economy*, 1989.
Pace University Center for Environmental Legal Studies, *Environment Costs of Electricity*, 1989.
R. W. Sant, *Creating Abundance: America's Least Cost Energy Strategy*, 1984.
Tellus Institute, *Full Economic Dispatch: Recognising Environmental Externalities in Electric Utility Systems Operation*, 1990.
S. Thomas, *Evaluation of RWE's Kes S Rebate Programme for Efficient Residential Appliances*, 1995.

Subject Index